Smell: A Very Short Introduction

VERY SHORT INTRODUCTIONS are for anyone wanting a stimulating and accessible way into a new subject. They are written by experts, and have been translated into more than 45 different languages.

The series began in 1995, and now covers a wide variety of topics in every discipline. The VSI library currently contains over 650 volumes—a Very Short Introduction to everything from Psychology and Philosophy of Science to American History and Relativity—and continues to grow in every subject area.

Very Short Introductions available now:

ABOLITIONISM Richard S. Newman
THE ABRAHAMIC RELIGIONS Charles L. Cohen
ACCOUNTING Christopher Nobes
ADAM SMITH Christopher J. Berry
ADOLESCENCE Peter K. Smith
ADVERTISING Winston Fletcher
AERIAL WARFARE Frank Ledwidge
AESTHETICS Bence Nanay
AFRICAN AMERICAN RELIGION Eddie S. Glaude Jr
AFRICAN HISTORY John Parker and Richard Rathbone
AFRICAN POLITICS Ian Taylor
AFRICAN RELIGIONS Jacob K. Olupona
AGEING Nancy A. Pachana
AGNOSTICISM Robin Le Poidevin
AGRICULTURE Paul Brassley and Richard Soffe
ALBERT CAMUS Oliver Gloag
ALEXANDER THE GREAT Hugh Bowden
ALGEBRA Peter M. Higgins
AMERICAN CULTURAL HISTORY Eric Avila
AMERICAN FOREIGN RELATIONS Andrew Preston
AMERICAN HISTORY Paul S. Boyer
AMERICAN IMMIGRATION David A. Gerber
AMERICAN LEGAL HISTORY G. Edward White
AMERICAN NAVAL HISTORY Craig L. Symonds
AMERICAN POLITICAL HISTORY Donald Critchlow
AMERICAN POLITICAL PARTIES AND ELECTIONS L. Sandy Maisel
AMERICAN POLITICS Richard M. Valelly
THE AMERICAN PRESIDENCY Charles O. Jones
THE AMERICAN REVOLUTION Robert J. Allison
AMERICAN SLAVERY Heather Andrea Williams
THE AMERICAN WEST Stephen Aron
AMERICAN WOMEN'S HISTORY Susan Ware
ANAESTHESIA Aidan O'Donnell
ANALYTIC PHILOSOPHY Michael Beaney
ANARCHISM Colin Ward
ANCIENT ASSYRIA Karen Radner
ANCIENT EGYPT Ian Shaw
ANCIENT EGYPTIAN ART AND ARCHITECTURE Christina Riggs
ANCIENT GREECE Paul Cartledge
THE ANCIENT NEAR EAST Amanda H. Podany
ANCIENT PHILOSOPHY Julia Annas
ANCIENT WARFARE Harry Sidebottom
ANGELS David Albert Jones
ANGLICANISM Mark Chapman
THE ANGLO-SAXON AGE John Blair
ANIMAL BEHAVIOUR Tristram D. Wyatt

THE ANIMAL KINGDOM
Peter Holland
ANIMAL RIGHTS David DeGrazia
THE ANTARCTIC Klaus Dodds
ANTHROPOCENE Erle C. Ellis
ANTISEMITISM Steven Beller
ANXIETY Daniel Freeman and
Jason Freeman
THE APOCRYPHAL GOSPELS
Paul Foster
APPLIED MATHEMATICS
Alain Goriely
ARCHAEOLOGY Paul Bahn
ARCHITECTURE Andrew Ballantyne
ARISTOCRACY William Doyle
ARISTOTLE Jonathan Barnes
ART HISTORY Dana Arnold
ART THEORY Cynthia Freeland
ARTIFICIAL INTELLIGENCE
Margaret A. Boden
ASIAN AMERICAN HISTORY
Madeline Y. Hsu
ASTROBIOLOGY David C. Catling
ASTROPHYSICS James Binney
ATHEISM Julian Baggini
THE ATMOSPHERE Paul I. Palmer
AUGUSTINE Henry Chadwick
AUSTRALIA Kenneth Morgan
AUTISM Uta Frith
AUTOBIOGRAPHY Laura Marcus
THE AVANT GARDE David Cottington
THE AZTECS Davíd Carrasco
BABYLONIA Trevor Bryce
BACTERIA Sebastian G. B. Amyes
BANKING John Goddard and
John O. S. Wilson
BARTHES Jonathan Culler
THE BEATS David Sterritt
BEAUTY Roger Scruton
BEHAVIOURAL ECONOMICS
Michelle Baddeley
BESTSELLERS John Sutherland
THE BIBLE John Riches
BIBLICAL ARCHAEOLOGY
Eric H. Cline
BIG DATA Dawn E. Holmes
BIOGRAPHY Hermione Lee
BIOMETRICS Michael Fairhurst
BLACK HOLES Katherine Blundell
BLOOD Chris Cooper
THE BLUES Elijah Wald
THE BODY Chris Shilling
THE BOOK OF COMMON PRAYER
Brian Cummings
THE BOOK OF MORMON
Terryl Givens
BORDERS Alexander C. Diener and
Joshua Hagen
THE BRAIN Michael O'Shea
BRANDING Robert Jones
THE BRICS Andrew F. Cooper
THE BRITISH CONSTITUTION
Martin Loughlin
THE BRITISH EMPIRE Ashley Jackson
BRITISH POLITICS Anthony Wright
BUDDHA Michael Carrithers
BUDDHISM Damien Keown
BUDDHIST ETHICS Damien Keown
BYZANTIUM Peter Sarris
C. S. LEWIS James Como
CALVINISM Jon Balserak
CANCER Nicholas James
CAPITALISM James Fulcher
CATHOLICISM Gerald O'Collins
CAUSATION Stephen Mumford and
Rani Lill Anjum
THE CELL Terence Allen and
Graham Cowling
THE CELTS BarryCunliffe
CHAOS Leonard Smith
CHARLES DICKENS Jenny Hartley
CHEMISTRY Peter Atkins
CHILD PSYCHOLOGY Usha Goswami
CHILDREN'S LITERATURE
Kimberley Reynolds
CHINESE LITERATURE Sabina Knight
CHOICE THEORY Michael Allingham
CHRISTIAN ART Beth Williamson
CHRISTIAN ETHICS D. Stephen Long
CHRISTIANITY Linda Woodhead
CIRCADIAN RHYTHMS
Russell Foster and Leon Kreitzman
CITIZENSHIP Richard Bellamy
CIVIL ENGINEERING
David Muir Wood
CLASSICAL LITERATURE William Allan
CLASSICAL MYTHOLOGY
Helen Morales
CLASSICS Mary Beard and
John Henderson
CLAUSEWITZ Michael Howard
CLIMATE Mark Maslin

CLIMATE CHANGE Mark Maslin
CLINICAL PSYCHOLOGY
Susan Llewelyn and
Katie Aafjes-van Doorn
COGNITIVE NEUROSCIENCE
Richard Passingham
THE COLD WAR Robert McMahon
COLONIAL AMERICA Alan Taylor
COLONIAL LATIN AMERICAN
LITERATURE Rolena Adorno
COMBINATORICS Robin Wilson
COMEDY Matthew Bevis
COMMUNISM Leslie Holmes
COMPARATIVE LITERATURE
Ben Hutchinson
COMPLEXITY John H. Holland
THE COMPUTER Darrel Ince
COMPUTER SCIENCE
Subrata Dasgupta
CONCENTRATION CAMPS
Dan Stone
CONFUCIANISM Daniel K. Gardner
THE CONQUISTADORS
Matthew Restall and
Felipe Fernández-Armesto
CONSCIENCE Paul Strohm
CONSCIOUSNESS Susan Blackmore
CONTEMPORARY ART
Julian Stallabrass
CONTEMPORARY FICTION
Robert Eaglestone
CONTINENTAL PHILOSOPHY
Simon Critchley
COPERNICUS Owen Gingerich
CORAL REEFS Charles Sheppard
CORPORATE SOCIAL
RESPONSIBILITY Jeremy Moon
CORRUPTION Leslie Holmes
COSMOLOGY Peter Coles
COUNTRY MUSIC Richard Carlin
CRIME FICTION Richard Bradford
CRIMINAL JUSTICE Julian V. Roberts
CRIMINOLOGY Tim Newburn
CRITICAL THEORY
Stephen Eric Bronner
THE CRUSADES Christopher Tyerman
CRYPTOGRAPHY Fred Piper and
Sean Murphy
CRYSTALLOGRAPHY A. M. Glazer
THE CULTURAL REVOLUTION
Richard Curt Kraus
DADA AND SURREALISM
David Hopkins
DANTE Peter Hainsworth and
David Robey
DARWIN Jonathan Howard
THE DEAD SEA SCROLLS
Timothy H. Lim
DECADENCE David Weir
DECOLONIZATION Dane Kennedy
DEMOCRACY Bernard Crick
DEMOGRAPHY Sarah Harper
DEPRESSION Jan Scott and
Mary Jane Tacchi
DERRIDA Simon Glendinning
DESCARTES Tom Sorell
DESERTS Nick Middleton
DESIGN John Heskett
DEVELOPMENT Ian Goldin
DEVELOPMENTAL BIOLOGY
Lewis Wolpert
THE DEVIL Darren Oldridge
DIASPORA Kevin Kenny
DICTIONARIES Lynda Mugglestone
DINOSAURS David Norman
DIPLOMACY Joseph M. Siracusa
DOCUMENTARY FILM
Patricia Aufderheide
DREAMING J. Allan Hobson
DRUGS Les Iversen
DRUIDS Barry Cunliffe
DYNASTY Jeroen Duindam
DYSLEXIA Margaret J. Snowling
EARLY MUSIC Thomas Forrest Kelly
THE EARTH Martin Redfern
EARTH SYSTEM SCIENCE Tim Lenton
ECONOMICS Partha Dasgupta
EDUCATION Gary Thomas
EGYPTIAN MYTH Geraldine Pinch
EIGHTEENTH-CENTURY BRITAIN
Paul Langford
THE ELEMENTS Philip Ball
EMOTION Dylan Evans
EMPIRE Stephen Howe
ENERGY SYSTEMS Nick Jenkins
ENGELS Terrell Carver
ENGINEERING David Blockley
THE ENGLISH LANGUAGE
Simon Horobin
ENGLISH LITERATURE Jonathan Bate
THE ENLIGHTENMENT
John Robertson

ENTREPRENEURSHIP Paul Westhead and Mike Wright
ENVIRONMENTAL ECONOMICS Stephen Smith
ENVIRONMENTAL ETHICS Robin Attfield
ENVIRONMENTAL LAW Elizabeth Fisher
ENVIRONMENTAL POLITICS Andrew Dobson
EPICUREANISM Catherine Wilson
EPIDEMIOLOGY Rodolfo Saracci
ETHICS Simon Blackburn
ETHNOMUSICOLOGY Timothy Rice
THE ETRUSCANS Christopher Smith
EUGENICS Philippa Levine
THE EUROPEAN UNION Simon Usherwood and John Pinder
EUROPEAN UNION LAW Anthony Arnull
EVOLUTION Brian and Deborah Charlesworth
EXISTENTIALISM Thomas Flynn
EXPLORATION Stewart A. Weaver
EXTINCTION Paul B. Wignall
THE EYE Michael Land
FAIRY TALE Marina Warner
FAMILY LAW Jonathan Herring
FASCISM Kevin Passmore
FASHION Rebecca Arnold
FEDERALISM Mark J. Rozell and Clyde Wilcox
FEMINISM Margaret Walters
FILM Michael Wood
FILM MUSIC Kathryn Kalinak
FILM NOIR James Naremore
THE FIRST WORLD WAR Michael Howard
FOLK MUSIC Mark Slobin
FOOD John Krebs
FORENSIC PSYCHOLOGY David Canter
FORENSIC SCIENCE Jim Fraser
FORESTS Jaboury Ghazoul
FOSSILS Keith Thomson
FOUCAULT Gary Gutting
THE FOUNDING FATHERS R. B. Bernstein
FRACTALS Kenneth Falconer
FREE SPEECH Nigel Warburton
FREE WILL Thomas Pink
FREEMASONRY Andreas Önnerfors
FRENCH LITERATURE John D. Lyons
THE FRENCH REVOLUTION William Doyle
FREUD Anthony Storr
FUNDAMENTALISM Malise Ruthven
FUNGI Nicholas P. Money
THE FUTURE Jennifer M. Gidley
GALAXIES John Gribbin
GALILEO Stillman Drake
GAME THEORY Ken Binmore
GANDHI Bhikhu Parekh
GARDEN HISTORY Gordon Campbell
GENES Jonathan Slack
GENIUS Andrew Robinson
GENOMICS John Archibald
GEOFFREY CHAUCER David Wallace
GEOGRAPHY John Matthews and David Herbert
GEOLOGY Jan Zalasiewicz
GEOPHYSICS William Lowrie
GEOPOLITICS Klaus Dodds
GERMAN LITERATURE Nicholas Boyle
GERMAN PHILOSOPHY Andrew Bowie
GLACIATION David J. A. Evans
GLOBAL CATASTROPHES Bill McGuire
GLOBAL ECONOMIC HISTORY Robert C. Allen
GLOBALIZATION Manfred Steger
GOD John Bowker
GOETHE Ritchie Robertson
THE GOTHIC Nick Groom
GOVERNANCE Mark Bevir
GRAVITY Timothy Clifton
THE GREAT DEPRESSION AND THE NEW DEAL Eric Rauchway
HABERMAS James Gordon Finlayson
THE HABSBURG EMPIRE Martyn Rady
HAPPINESS Daniel M. Haybron
THE HARLEM RENAISSANCE Cheryl A. Wall
THE HEBREW BIBLE AS LITERATURE Tod Linafelt
HEGEL Peter Singer
HEIDEGGER Michael Inwood
THE HELLENISTIC AGE Peter Thonemann
HEREDITY John Waller
HERMENEUTICS Jens Zimmermann

HERODOTUS Jennifer T. Roberts
HIEROGLYPHS Penelope Wilson
HINDUISM Kim Knott
HISTORY John H. Arnold
THE HISTORY OF ASTRONOMY Michael Hoskin
THE HISTORY OF CHEMISTRY William H. Brock
THE HISTORY OF CHILDHOOD James Marten
THE HISTORY OF CINEMA Geoffrey Nowell-Smith
THE HISTORY OF LIFE Michael Benton
THE HISTORY OF MATHEMATICS Jacqueline Stedall
THE HISTORY OF MEDICINE William Bynum
THE HISTORY OF PHYSICS J. L. Heilbron
THE HISTORY OF TIME Leofranc Holford-Strevens
HIV AND AIDS Alan Whiteside
HOBBES Richard Tuck
HOLLYWOOD Peter Decherney
THE HOLY ROMAN EMPIRE Joachim Whaley
HOME Michael Allen Fox
HOMER Barbara Graziosi
HORMONES Martin Luck
HUMAN ANATOMY Leslie Klenerman
HUMAN EVOLUTION Bernard Wood
HUMAN RIGHTS Andrew Clapham
HUMANISM Stephen Law
HUME A. J. Ayer
HUMOUR Noël Carroll
THE ICE AGE Jamie Woodward
IDENTITY Florian Coulmas
IDEOLOGY Michael Freeden
THE IMMUNE SYSTEM Paul Klenerman
INDIAN CINEMA Ashish Rajadhyaksha
INDIAN PHILOSOPHY Sue Hamilton
THE INDUSTRIAL REVOLUTION Robert C. Allen
INFECTIOUS DISEASE Marta L. Wayne and Benjamin M. Bolker
INFINITY Ian Stewart
INFORMATION Luciano Floridi
INNOVATION Mark Dodgson and David Gann
INTELLECTUAL PROPERTY Siva Vaidhyanathan
INTELLIGENCE Ian J. Deary
INTERNATIONAL LAW Vaughan Lowe
INTERNATIONAL MIGRATION Khalid Koser
INTERNATIONAL RELATIONS Christian Reus-Smit
INTERNATIONAL SECURITY Christopher S. Browning
IRAN Ali M. Ansari
ISLAM Malise Ruthven
ISLAMIC HISTORY Adam Silverstein
ISOTOPES Rob Ellam
ITALIAN LITERATURE Peter Hainsworth and David Robey
JESUS Richard Bauckham
JEWISH HISTORY David N. Myers
JOURNALISM Ian Hargreaves
JUDAISM Norman Solomon
JUNG Anthony Stevens
KABBALAH Joseph Dan
KAFKA Ritchie Robertson
KANT Roger Scruton
KEYNES Robert Skidelsky
KIERKEGAARD Patrick Gardiner
KNOWLEDGE Jennifer Nagel
THE KORAN Michael Cook
KOREA Michael J. Seth
LAKES Warwick F. Vincent
LANDSCAPE ARCHITECTURE Ian H. Thompson
LANDSCAPES AND GEOMORPHOLOGY Andrew Goudie and Heather Viles
LANGUAGES Stephen R. Anderson
LATE ANTIQUITY Gillian Clark
LAW Raymond Wacks
THE LAWS OF THERMODYNAMICS Peter Atkins
LEADERSHIP Keith Grint
LEARNING Mark Haselgrove
LEIBNIZ Maria Rosa Antognazza
LEO TOLSTOY Liza Knapp
LIBERALISM Michael Freeden
LIGHT Ian Walmsley
LINCOLN Allen C. Guelzo
LINGUISTICS Peter Matthews

LITERARY THEORY Jonathan Culler
LOCKE John Dunn
LOGIC Graham Priest
LOVE Ronald de Sousa
MACHIAVELLI Quentin Skinner
MADNESS Andrew Scull
MAGIC Owen Davies
MAGNA CARTA Nicholas Vincent
MAGNETISM Stephen Blundell
MALTHUS Donald Winch
MAMMALS T. S. Kemp
MANAGEMENT John Hendry
MAO Delia Davin
MARINE BIOLOGY Philip V. Mladenov
THE MARQUIS DE SADE John Phillips
MARTIN LUTHER Scott H. Hendrix
MARTYRDOM Jolyon Mitchell
MARX Peter Singer
MATERIALS Christopher Hall
MATHEMATICAL FINANCE Mark H. A. Davis
MATHEMATICS Timothy Gowers
MATTER Geoff Cottrell
THE MEANING OF LIFE Terry Eagleton
MEASUREMENT David Hand
MEDICAL ETHICS Michael Dunn and Tony Hope
MEDICAL LAW Charles Foster
MEDIEVAL BRITAIN John Gillingham and Ralph A. Griffiths
MEDIEVAL LITERATURE Elaine Treharne
MEDIEVAL PHILOSOPHY John Marenbon
MEMORY Jonathan K. Foster
METAPHYSICS Stephen Mumford
METHODISM William J. Abraham
THE MEXICAN REVOLUTION Alan Knight
MICHAEL FARADAY Frank A. J. L. James
MICROBIOLOGY Nicholas P. Money
MICROECONOMICS Avinash Dixit
MICROSCOPY Terence Allen
THE MIDDLE AGES Miri Rubin
MILITARY JUSTICE Eugene R. Fidell
MILITARY STRATEGY Antulio J. Echevarria II
MINERALS David Vaughan
MIRACLES Yujin Nagasawa
MODERN ARCHITECTURE Adam Sharr
MODERN ART David Cottington
MODERN CHINA Rana Mitter
MODERN DRAMA Kirsten E. Shepherd-Barr
MODERN FRANCE Vanessa R. Schwartz
MODERN INDIA Craig Jeffrey
MODERN IRELAND Senia Pašeta
MODERN ITALY Anna Cento Bull
MODERN JAPAN Christopher Goto-Jones
MODERN LATIN AMERICAN LITERATURE Roberto González Echevarría
MODERN WAR Richard English
MODERNISM Christopher Butler
MOLECULAR BIOLOGY Aysha Divan and Janice A. Royds
MOLECULES Philip Ball
MONASTICISM Stephen J. Davis
THE MONGOLS Morris Rossabi
MOONS David A. Rothery
MORMONISM Richard Lyman Bushman
MOUNTAINS Martin F. Price
MUHAMMAD Jonathan A. C. Brown
MULTICULTURALISM Ali Rattansi
MULTILINGUALISM John C. Maher
MUSIC Nicholas Cook
MYTH Robert A. Segal
NAPOLEON David Bell
THE NAPOLEONIC WARS Mike Rapport
NATIONALISM Steven Grosby
NATIVE AMERICAN LITERATURE Sean Teuton
NAVIGATION Jim Bennett
NAZI GERMANY Jane Caplan
NELSON MANDELA Elleke Boehmer
NEOLIBERALISM Manfred Steger and Ravi Roy
NETWORKS Guido Caldarelli and Michele Catanzaro
THE NEW TESTAMENT Luke Timothy Johnson
THE NEW TESTAMENT AS LITERATURE Kyle Keefer
NEWTON Robert Iliffe
NIELS BOHR J. L. Heilbron

NIETZSCHE Michael Tanner
NINETEENTH-CENTURY BRITAIN
Christopher Harvie and
H. C. G. Matthew
THE NORMAN CONQUEST
George Garnett
NORTH AMERICAN INDIANS
Theda Perdue and Michael D. Green
NORTHERN IRELAND
Marc Mulholland
NOTHING Frank Close
NUCLEAR PHYSICS Frank Close
NUCLEAR POWER Maxwell Irvine
NUCLEAR WEAPONS
Joseph M. Siracusa
NUMBER THEORY Robin Wilson
NUMBERS Peter M. Higgins
NUTRITION David A. Bender
OBJECTIVITY Stephen Gaukroger
OCEANS Dorrik Stow
THE OLD TESTAMENT
Michael D. Coogan
THE ORCHESTRA D. Kern Holoman
ORGANIC CHEMISTRY
Graham Patrick
ORGANIZATIONS Mary Jo Hatch
ORGANIZED CRIME
Georgios A. Antonopoulos and
Georgios Papanicolaou
ORTHODOX CHRISTIANITY
A. Edward Siecienski
PAGANISM Owen Davies
PAIN Rob Boddice
THE PALESTINIAN-ISRAELI
CONFLICT Martin Bunton
PANDEMICS Christian W. McMillen
PARTICLE PHYSICS Frank Close
PAUL E. P. Sanders
PEACE Oliver P. Richmond
PENTECOSTALISM William K. Kay
PERCEPTION Brian Rogers
THE PERIODIC TABLE Eric R. Scerri
PHILOSOPHY Edward Craig
PHILOSOPHY IN THE ISLAMIC
WORLD Peter Adamson
PHILOSOPHY OF BIOLOGY
Samir Okasha
PHILOSOPHY OF LAW
Raymond Wacks
PHILOSOPHY OF SCIENCE
Samir Okasha
PHILOSOPHY OF RELIGION
Tim Bayne
PHOTOGRAPHY Steve Edwards
PHYSICAL CHEMISTRY Peter Atkins
PHYSICS Sidney Perkowitz
PILGRIMAGE Ian Reader
PLAGUE Paul Slack
PLANETS David A. Rothery
PLANTS Timothy Walker
PLATE TECTONICS Peter Molnar
PLATO Julia Annas
POETRY Bernard O'Donoghue
POLITICAL PHILOSOPHY
David Miller
POLITICS Kenneth Minogue
POPULISM Cas Mudde and
Cristóbal Rovira Kaltwasser
POSTCOLONIALISM Robert Young
POSTMODERNISM Christopher Butler
POSTSTRUCTURALISM
Catherine Belsey
POVERTY Philip N. Jefferson
PREHISTORY Chris Gosden
PRESOCRATIC PHILOSOPHY
Catherine Osborne
PRIVACY Raymond Wacks
PROBABILITY John Haigh
PROGRESSIVISM Walter Nugent
PROHIBITION W. J. Rorabaugh
PROJECTS Andrew Davies
PROTESTANTISM Mark A. Noll
PSYCHIATRY Tom Burns
PSYCHOANALYSIS Daniel Pick
PSYCHOLOGY Gillian Butler and
Freda McManus
PSYCHOLOGY OF MUSIC
Elizabeth Hellmuth Margulis
PSYCHOPATHY Essi Viding
PSYCHOTHERAPY Tom Burns and
Eva Burns-Lundgren
PUBLIC ADMINISTRATION
Stella Z. Theodoulou and Ravi K. Roy
PUBLIC HEALTH Virginia Berridge
PURITANISM Francis J. Bremer
THE QUAKERS Pink Dandelion
QUANTUM THEORY
John Polkinghorne
RACISM Ali Rattansi
RADIOACTIVITY Claudio Tuniz
RASTAFARI Ennis B. Edmonds
READING Belinda Jack

THE REAGAN REVOLUTION Gil Troy
REALITY Jan Westerhoff
RECONSTRUCTION Allen C. Guelzo
THE REFORMATION Peter Marshall
RELATIVITY Russell Stannard
RELIGION IN AMERICA Timothy Beal
THE RENAISSANCE Jerry Brotton
RENAISSANCE ART Geraldine A. Johnson
RENEWABLE ENERGY Nick Jelley
REPTILES T. S. Kemp
REVOLUTIONS Jack A. Goldstone
RHETORIC Richard Toye
RISK Baruch Fischhoff and John Kadvany
RITUAL Barry Stephenson
RIVERS Nick Middleton
ROBOTICS Alan Winfield
ROCKS Jan Zalasiewicz
ROMAN BRITAIN Peter Salway
THE ROMAN EMPIRE Christopher Kelly
THE ROMAN REPUBLIC David M. Gwynn
ROMANTICISM Michael Ferber
ROUSSEAU Robert Wokler
RUSSELL A. C. Grayling
RUSSIAN HISTORY Geoffrey Hosking
RUSSIAN LITERATURE Catriona Kelly
THE RUSSIAN REVOLUTION S. A. Smith
SAINTS Simon Yarrow
SAVANNAS Peter A. Furley
SCEPTICISM Duncan Pritchard
SCHIZOPHRENIA Chris Frith and Eve Johnstone
SCHOPENHAUER Christopher Janaway
SCIENCE AND RELIGION Thomas Dixon
SCIENCE FICTION David Seed
THE SCIENTIFIC REVOLUTION Lawrence M. Principe
SCOTLAND Rab Houston
SECULARISM Andrew Copson
SEXUAL SELECTION Marlene Zuk and Leigh W. Simmons
SEXUALITY Véronique Mottier
SHAKESPEARE'S COMEDIES Bart van Es
SHAKESPEARE'S SONNETS AND POEMS Jonathan F. S. Post
SHAKESPEARE'S TRAGEDIES Stanley Wells
SIKHISM Eleanor Nesbitt
THE SILK ROAD James A. Millward
SLANG Jonathon Green
SLEEP Steven W. Lockley and Russell G. Foster
SMELL Matthew Cobb
SOCIAL AND CULTURAL ANTHROPOLOGY John Monaghan and Peter Just
SOCIAL PSYCHOLOGY Richard J. Crisp
SOCIAL WORK Sally Holland and Jonathan Scourfield
SOCIALISM Michael Newman
SOCIOLINGUISTICS John Edwards
SOCIOLOGY Steve Bruce
SOCRATES C. C. W. Taylor
SOUND Mike Goldsmith
SOUTHEAST ASIA James R. Rush
THE SOVIET UNION Stephen Lovell
THE SPANISH CIVIL WAR Helen Graham
SPANISH LITERATURE Jo Labanyi
SPINOZA Roger Scruton
SPIRITUALITY Philip Sheldrake
SPORT Mike Cronin
STARS Andrew King
STATISTICS David J. Hand
STEM CELLS Jonathan Slack
STOICISM Brad Inwood
STRUCTURAL ENGINEERING David Blockley
STUART BRITAIN John Morrill
SUPERCONDUCTIVITY Stephen Blundell
SUPERSTITION Stuart Vyse
SYMMETRY Ian Stewart
SYNAESTHESIA Julia Simner
SYNTHETIC BIOLOGY Jamie A. Davies
SYSTEMS BIOLOGY Eberhard O. Voit
TAXATION Stephen Smith
TEETH Peter S. Ungar
TELESCOPES Geoff Cottrell
TERRORISM Charles Townshend
THEATRE Marvin Carlson
THEOLOGY David F. Ford
THINKING AND REASONING Jonathan St B. T. Evans
THOMAS AQUINAS Fergus Kerr
THOUGHT Tim Bayne

TIBETAN BUDDHISM Matthew T. Kapstein
TIDES David George Bowers and Emyr Martyn Roberts
TOCQUEVILLE Harvey C. Mansfield
TOPOLOGY Richard Earl
TRAGEDY Adrian Poole
TRANSLATION Matthew Reynolds
THE TREATY OF VERSAILLES Michael S. Neiberg
TRIGONOMETRY Glen Van Brummelen
THE TROJAN WAR Eric H. Cline
TRUST Katherine Hawley
THE TUDORS John Guy
TWENTIETH-CENTURY BRITAIN Kenneth O. Morgan
TYPOGRAPHY Paul Luna
THE UNITED NATIONS Jussi M. Hanhimäki
UNIVERSITIES AND COLLEGES David Palfreyman and Paul Temple
THE U.S. CONGRESS Donald A. Ritchie
THE U.S. CONSTITUTION David J. Bodenhamer
THE U.S. SUPREME COURT Linda Greenhouse
UTILITARIANISM Katarzyna de Lazari-Radek and Peter Singer
UTOPIANISM Lyman Tower Sargent
VETERINARY SCIENCE James Yeates
THE VIKINGS Julian D. Richards
VIRUSES Dorothy H. Crawford
VOLTAIRE Nicholas Cronk
WAR AND TECHNOLOGY Alex Roland
WATER John Finney
WAVES Mike Goldsmith
WEATHER Storm Dunlop
THE WELFARE STATE David Garland
WILLIAM SHAKESPEARE Stanley Wells
WITCHCRAFT Malcolm Gaskill
WITTGENSTEIN A. C. Grayling
WORK Stephen Fineman
WORLD MUSIC Philip Bohlman
THE WORLD TRADE ORGANIZATION Amrita Narlikar
WORLD WAR II Gerhard L. Weinberg
WRITING AND SCRIPT Andrew Robinson
ZIONISM Michael Stanislawski

Available soon:

THE SUN Philip Judge
FIRE Andrew C. Scott
ECOLOGY Jaboury Ghazoul
MODERN BRAZIL Anthony W. Pereira
BIOGEOGRAPHY Mark V. Lomolino

For more information visit our website

www.oup.com/vsi/

Matthew Cobb

A Very Short Introduction

OXFORD
UNIVERSITY PRESS

Great Clarendon Street, Oxford, OX2 6DP,
United Kingdom

Oxford University Press is a department of the University of Oxford.
It furthers the University's objective of excellence in research, scholarship,
and education by publishing worldwide. Oxford is a registered trade mark of
Oxford University Press in the UK and in certain other countries

First edition published in 2020

Impression: 1

Published in the United States of America by Oxford University Press
198 Madison Avenue, New York, NY 10016, United States of America

British Library Cataloguing in Publication Data
Data available

Library of Congress Control Number: 2020934672

ISBN 978–0–19–882525–8

Printed in Great Britain by
Ashford Colour Press Ltd, Gosport, Hampshire

Contents

Acknowledgements

I have been studying smell, mainly in insects, for over thirty years—first at the Centre National de la Recherche Scientifique in the Paris region and then, since 2002, at the University of Manchester. In that time, many colleagues and friends around the world have taught me so much—in the laboratory, in conferences, in their books and scientific articles, and in informal discussions. Their collective influence can be felt in the pages that follow. Particular thanks go to Tristram Wyatt and Kara Hoover for detailed comments on the manuscript, and to Catherine McCrohan, Krishna Persaud, and Leslie Vosshall both for specific help and for many discussions over the years. My long research collaboration with Catherine McCrohan—an expert electrophysiologist—on the sense of smell in maggots has played a fundamental role in shaping how I think about the early detection of odours. Since 2004 I have taught a final year undergraduate course at Manchester entitled Chemical Communication in Animals, which has been taken by over 1,000 students. Those young people have kept me on my toes and provided an invaluable sounding-board for the ideas presented here. If any of you are reading this—thanks!

This book is jointly dedicated to two friends and colleagues: Jean-Marc Jallon (1945–2019), who pushed me into studying olfaction, much against my will, and Dr Victoria Henshaw (1971–2014), whose pioneering work on smell in urban environments was so inspirational. Both of them are sorely missed.

List of illustrations

Chapter 1
How we smell

What is your favourite smell? People often choose baking bread, freshly brewed coffee, or the zing of an orange. My own choices would be petrichor (the smell produced when rain has fallen on dry soil), the resiny tang of pine forests, or, above all, the comforting warm smell of the back of a baby's head. Sensing each of these smells brings back experiences associated with them, in a way that does not happen so richly for, say, vision or sound. Smell is special.

Smell—or olfaction, to give it its scientific name—is probably the oldest sense. Organisms were able to detect chemicals in their environment and directly respond to them long before they could see or feel, although the mechanisms involved have long since changed profoundly. Animals use smell for a range of essential functions—such as finding food, avoiding predators, locating shelter, obtaining a mate, or as a fundamental part of memory. To detect those smells, they use a variety of structures. Vertebrates bring odours into their body via their nostrils or mouth (Figure 1), while insects and crustaceans use antennae to capture odours (other animals such as worms or snails have specialized organs on their heads that serve the same function). In all cases, these organs are linked directly to the brain, enabling the animal to rapidly identify a smell and its location, and to respond appropriately to it.

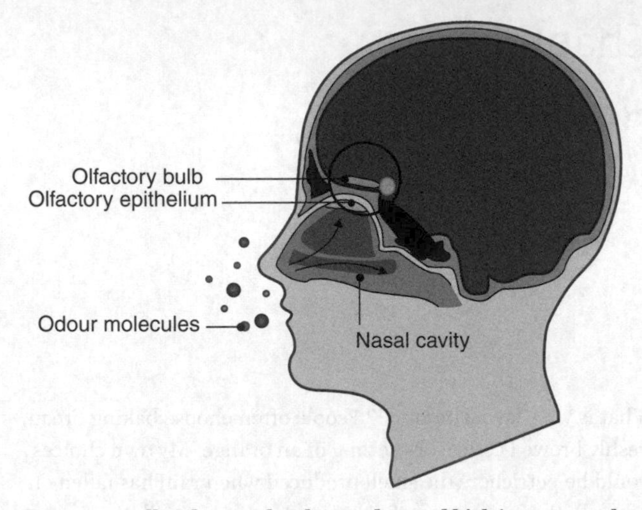

1. How you smell. Odour molecules are detected high in your nasal cavity, about the level of your eyes.

Despite being so fundamental for all animals, including us, the sense of smell remains mysterious. We understand far less about it than we do vision, touch, taste, or hearing. The enigma of olfaction can be highlighted by trying to describe a scent—the words we use generally refer to other smells, or, perhaps, to some multisensory term like 'green'. To put it technically—what is the dimensionality of smell? We can define visual and auditory stimuli in terms of wavelength and intensity. But odours resist verbal categorization. There are some exceptions to this in English ('musty', 'acrid', 'dank'...) but in general, we do not have words for smells. The best scientists can do is to classify odours—which are molecules—by their various component atoms, size, and structure, which seem to be linked with how we perceive the odour.

For example, the smell of rotting vegetable matter is largely due to the presence of sulphur atoms in the molecule (Figure 2). Sweets like pear drops smell the way they do because the key aroma they contain is a small molecule composed of carbon and hydrogen

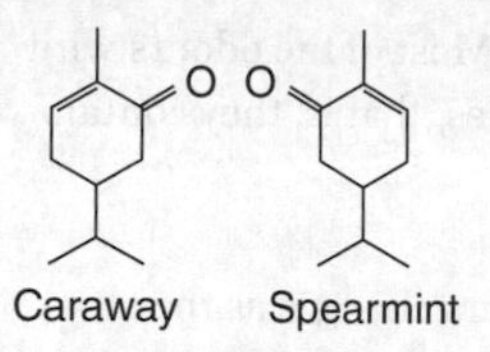

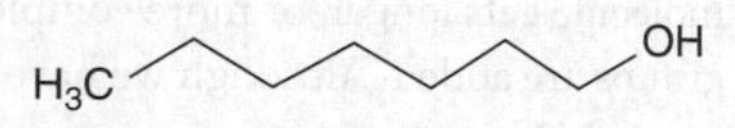

Heptanol - 'violet, sweet, woody'

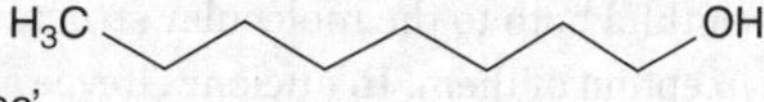

Octanol - 'sweet, orange, rose'

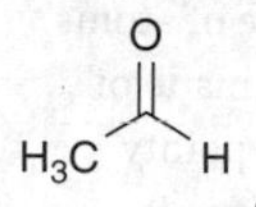

Acetaldehyde (fruity)

O

H_3C OH

Acetic acid (vinegar)

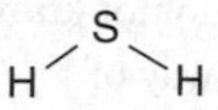

Hydrogen sulphide (rotting eggs)

S

H_3C CH_3

Dimethyl sulphide (rotten cabbage)

2. Molecular structures of various odours. Caraway and spearmint have the same structure but different orientations. The addition of a single atom of carbon can change our perception from 'violet, sweet, woody' (heptanol, C7) to 'sweet, orange, rose' (octanol, C8). A single atom of oxygen turns a fruity smell into vinegar (acetaldehyde and acetic acid), while additional carbons and hydrogens transform the smell of rotting eggs into the odour of rotten cabbage (hydrogen sulphide and dimethyl sulphide).

atoms forming what is called an ester. Esters and sulphurs are examples of functional groups, which affect both the chemical behaviour of the molecule and how we perceive it. These effects interact with the overall size or length of the molecule to give many smells their unique characteristics. Most of the odours with which we are familiar are organic molecules, that is, they contain carbon, which is omnipresent on Earth.

As Figure 2 shows, our perception of odours changes as the molecule gets longer, or more complex, or different functional groups are added. Although we have developed machines that can precisely identify the atomic composition of odours, we are unable, for the moment, to produce a systematic classification that is faithful both to the molecular structures of smells and to our perception of them. In ancient Greece and Rome, materialist philosophers such as Democritus and Lucretius argued that given all matter is made of atoms, then smells must be made of atoms too. They decided that things that smell nice must be made of round atoms, while sharp and acrid smells must have pointy atoms. While the detail is obviously wrong, the basic idea is correct—there is a link between the molecular structure of a smell and how we perceive it. What exactly that link is, we do not know. The dimensionality of smell—if that is even the right way of thinking about it—remains unknown.

The molecular limits of smell are not easy to define. Humans perceive odours that are present in the air; this requires molecules to be volatile, which is a function of the size of the molecule (if it is too heavy, it cannot get airborne) and of both temperature and humidity (higher levels of both these factors increase the likelihood of an odour being released from its substrate). It is often argued that molecules consisting of more than twenty-three carbons are not volatile, but this figure is not an absolute threshold and temperature and pressure will alter the exact point at which an odour takes to the air. Furthermore, even on the stillest of days, air currents will transport smell molecules. All this is different for smells in water, which need to be soluble and will slowly diffuse as well as being carried by the current.

Most studies of smell, whether in humans or other animals, are done in the laboratory, where brief pulses of single odours of known composition and concentration are presented to a subject and their responses are recorded. But the real world is not like a laboratory. Most scents that we smell in the real world are not single molecules but blends of different compounds. For example, a rose produces over 250 identifiable molecules in its fragrance, although that does not mean that when we smell a rose we are detecting all of them. In the case of the tomato, there are over 400 molecules in its bouquet, but only a couple of dozen are detectable by humans. The olfactory world is far richer than we imagine. Odours also come in different concentrations, and our feelings about an odour change with its intensity—imagine the difference between a dab of perfume behind the ears and spilling the whole bottle on the carpet. What is called the hedonic value of the odour—whether it is nice or nasty—may change with concentration, and yet Chanel No. 5 still smells like Chanel No. 5 in all cases. Working out how our brains process such everyday experiences is one of the challenges faced by olfactory neuroscience.

Four myths of smell

Because our grasp of smell is so poor, a number of myths and misunderstandings have accumulated around this sense. Clearing them up at the outset will help you on this journey to the centre of scent.

Myth 1: You smell with your nose. Although we can indeed smell by inhaling through our nostrils, we detect odours using neurons that are directly connected to the brain, which dangle down through the base of the skull, at about eye level (Figure 1). Really, you are smelling with your brain.

Myth 2: You smell molecules in the air. The molecules we smell are carried on the air, but we do not directly detect airborne odours. If your olfactory neurons were in contact with the air they would shrivel and die; instead, these neurons, which are found in a thin layer of skin called the olfactory epithelium, are protected by a layer of mucus. A similar thing happens with insects—their olfactory neurons bathe in liquid housed in their sensory hairs. For obvious reasons, aquatic organisms can only sense odours in water. (Yes, fish, crabs, lobsters, and so on all smell, too.) Any airborne molecule you want to smell has to get through that protective liquid barrier. For this to happen, there are special molecules called odorant binding proteins (OBPs) that are found in that mucus—effectively on the outside of your body. Their role seems to be to transport smell molecules through the mucus and deliver them to the receptors on the olfactory neurons.

Myth 3: We have a poor sense of smell. For a long time, scientists agreed with this claim, but in 2017, the neuroscientist John McGann of Rutgers University highlighted the real situation in an article entitled 'Poor human olfaction is a 19th-century myth'. He summarized the key evidence and concluded that 'human olfaction is excellent and impactful'. While it is generally the case that your sense of smell might be diminished if you are aged, or if you smoke, and you will generally be better at smelling if you are female (overall, women have a more acute sense of smell than men), in reality we all have an atomic nose (Figure 2). Your sense of smell can distinguish between molecules that differ in size by a single atom of carbon—people describe heptanol (an alcohol made of seven carbon atoms) as smelling 'violet, sweet, woody', while octanol (just one carbon atom more) smells 'sweet, orange, rose'. Caraway—one of the components of curry—smells different from spearmint; the two smells have exactly the same atomic composition, but their structure is different. The molecules are mirror images of each other, like two gloves. The difference in our perception of those two odours is due to the way our olfactory neurons respond to the different molecular orientations.

So fine is our sense of smell that the number of odours we can distinguish may be near-infinite. For decades, researchers repeated that the average human could distinguish about 10,000 odours, but this figure had no scientific basis. In 2014 researchers in the laboratory of my good friend Leslie Vosshall at Rockefeller University tried to estimate how many smells we might be able to tell apart, based on mixtures of molecules. They came up with the astonishing figure of over a trillion. Although this mathematical model has been challenged, it seems probable that there is no real limit to the number of odours we can detect. The same will apply to many other animals.

Myth 4: We do not use smell much. Your sense of smell and your sense of taste are intimately connected. If you try eating something tasty while pinching your nose, making sure your mouth is shut, you will find that there is little flavour; but when you take your fingers away, you should get a sudden rush of sensation as volatile compounds from the food you are chewing whoosh up your nasal cavity and flow over your olfactory neurons, high in your head. This is called retronasal olfaction and is one of the two ways you can detect odours; the other way is called orthonasal olfaction or, more simply, sniffing. Our sense of taste is relatively rudimentary, divided into a small number of classes (the traditional salt, sour, bitter, and sweet, together with the more recently identified umami (meaty), fatty, hot/spicy, metallic, and the taste of carbon dioxide), while our perception of flavour is a mixture of the simple world of taste and the rich, multiple dimensions of smell. In many languages, flavour is colloquially called 'taste' even though smell may be the dominant sense giving flavour its subtlety.

People who have no sense of smell (they are described as being anosmic) through injury, disease, or for genetic reasons, or simply because they are elderly, may find that their enjoyment of food is reduced because they cannot smell properly. They may worry excessively about leaving the gas on, or about their personal hygiene. Indeed, permanent loss of the sense of smell can lead to mental health problems, as once-familiar odours—the scent of loved ones, for example—are no longer perceptible. Disease and damage can also cause people to sense odours that are not actually there—this is known as phantosmia. This may be relatively benign (when I get a cold, I sometimes smell what I can only describe as weird toast), but in some cases it can be very disturbing, as people continually smell faeces or vomit, or even unnameable and unidentifiable smells.

These biological effects are melded with cultural norms over which smells are and are not socially acceptable. Over recent decades in the West, we have become very conscious of avoiding personal odours, and spend a great deal of money and effort trying to ensure we do not smell of sweat or other odours that are perceived as unpleasant. In other times and other societies, different rules have applied.

Smell has long been a powerful aspect of human culture. Fragrances have been used in rituals and ceremonies down the ages, while perfumes—often based on animal sources, such as musk from the scent glands of deer, or from plants—have been a significant part of many cultures. The modern perfume industry is perpetually creating new scents that promise glamour and excitement for both genders, in a way that other products cannot. Smell has even been used to tell the time—from the 11th century onwards, Chinese temples used aroma clocks, containing powdered incense that burned at a particular rate, to release different scents at different times.

Smell is not just a mysterious biological phenomenon, with fascinating examples from across the animal kingdom, it is also a key part of our social existence that often goes unrecognized. By understanding it better, using examples from a wide range of animals, we can gain insight not only into the natural world, but also its role in our culture. To do that, we first need to understand how exactly the sense of smell works. What follows is a bit technical, but I have done my best to keep it simple and to avoid unnecessary detail. When we encounter striking examples of human and animal smelling later on in the book, remember that they are all based on these processes. In some cases, our knowledge is so precise that we can explain those examples in terms of what is going on in our cells and our brains. This amazing machinery, which we do not fully understand, is at work in your head, and that of every other animal on the planet, right now.

The mechanics of smell

Each species of animal can detect a different range of odours. No species can detect all the molecules that are present in the environment in which it lives—there are some things that we cannot smell but which other terrestrial animals can, and vice versa. There are also differences between individuals, relating to the ability to smell an odour, or how pleasant it seems. For example, some people like the taste of coriander—known as cilantro in the USA—while others find it soapy and unpleasant; this effect has an underlying genetic component due to differences in the genes controlling our sense of smell (if you hate the taste of coriander, try eating it with your mouth closed while holding your nose—it won't taste soapy). Ultimately, the selection of scents detected by a given species, and how that odour is perceived, will depend upon the animal's ecology. The response profile of each species will enable it to locate sources of smell that are relevant to it and to respond accordingly.

Despite these differences between species, the way in which odours are processed is fundamentally similar in all animals, from the initial detection in the periphery right through to deciding how to respond. When a terrestrial animal first encounters an odour, the smell is absorbed by the mucus (in the case of vertebrates) or the fluid-filled interior of a sensory hair (in the case of an arthropod), as the odour molecule binds to an odorant binding protein (OBP). These proteins appear to act as molecular chaperones, transporting the odour to the receptor through the protective mucus or fluid. Their precise functions are still unclear, but their significance in the ecology of some animals is known. For example, in the fly *Drosophila sechellia*, the presence of one kind of OBP enables these insects to exploit the pungent fruit of the Indian mulberry (*Morinda citrifolia*) by detecting the plant's unique odour, something that competing species which lack this protein cannot do. In pandas, one particular OBP binds with odours that are found in bamboo, the animal's sole food. In some species, these proteins seem to do other things, such as responding to humidity or detecting chemical signals inside the body, as well as carrying out their chaperone function. Intriguingly, although humans have a number of genes that enable us to synthesize these proteins, only one kind of OBP has so far been detected in our olfactory epithelium. It has been suggested that small variations in this gene which lead to changes in the structure of the protein may explain inter-individual differences in olfactory sensitivity.

The protective fluid or mucus also contains enzymes that snip up smell molecules, removing them after they have been smelled, or even preventing some compounds from being smelled in the first place. For example, benzaldehyde is a small molecule that smells like almonds; mice appear not to be able to smell this odour because enzymes in the animal's mucus degrade it into two components—benzyl alcohol and benzoic acid. Similar things seem to happen in your nose—in our olfactory mucus the molecule hexanal gets turned into hexanol, suggesting that we cannot actually smell hexanal. Our sense of smell—like our other senses—is filtered and selected even before it gets to our brains.

In vertebrates, olfactory neurons are located in the olfactory epithelium, a layer of skin on the underside of the skull that is effectively divided into two halves, each associated with one of your nasal cavities, and projecting to one half of the brain

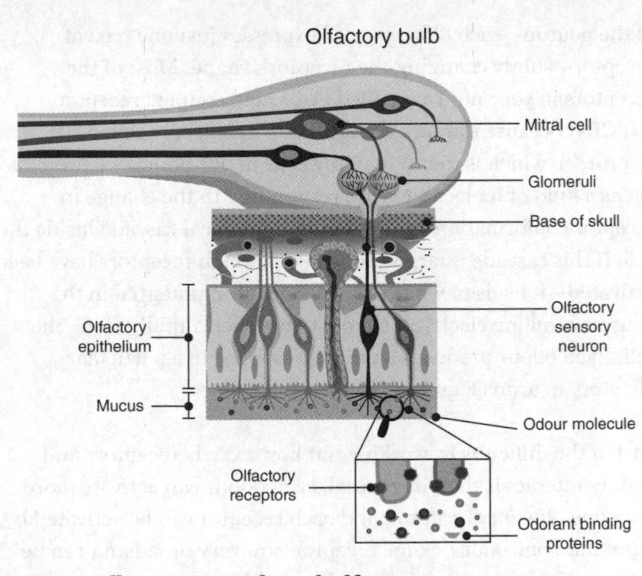

3. How smells are processed—each olfactory sensory neuron expresses a single type of olfactory receptor; all neurons of the same kind go to the same glomerulus in the brain's olfactory bulb.

(Figure 3). In some species, such as sheep or dogs, the olfactory epithelium can be relatively large, because the nasal cavity contains very convoluted bony structures called turbinates which are covered with the olfactory epithelium. This large surface area increases the sensitivity of the olfactory system by increasing the number of receptors. In a mouse the olfactory epithelium is rolled up into a tiny space; unravelled, it covers about 140 mm^2, or an area about the size of the top half of the mouse's head. There are supposedly about 10 million neurons packed onto that surface. In your nose the epithelium covers about 900 mm^2 (six times larger than a mouse's)—there are as yet no reliable estimates for the number of human olfactory neurons, but the number clearly runs into millions.

Once an odour molecule has arrived at the olfactory neuron, it binds to one of many millions of receptor proteins on the surface of the neuron—each olfactory cell expresses just one type of receptor—subtly changing the receptor's shape. Most of the receptors in your nose are called G-protein coupled receptors (GPCRs) because they are connected to a molecule called a G-protein, which is found in many cells in our bodies. G-proteins act as a kind of molecular relay, responding to the change in receptor conformation by producing a chemical cascade inside the cell. If this cascade is large enough—if enough receptors have been activated—it leads to what is called an action potential in the neuron, a rolling electrical charge that travels rapidly along the cell. Each odour produces a unique response in a particular olfactory neuron (Figure 4).

Part of the difficulty in working out how exactly receptors and odours interact is that, in general, each odour can activate more than one olfactory receptor, and each receptor can be activated by more than one odour. Some receptors are very broad and can be activated by a large number of odours; others are much more narrowly tuned and respond only to a small number of ecologically significant smells, or, particularly in the case of pheromones—chemical signals between members of the same species—sometimes to just one kind of odour molecule.

The complex pattern of activity in the periphery of our olfactory system is called a combinatorial code—each odour produces a unique pattern of neuronal activation of a specific subset of olfactory neurons, which is the basis of our perception of different odours (Figure 5). The olfactory combinatorial code is particularly complex because neurons do not respond in a digital manner—on or off. Each neuron can be excited or inhibited by different odours, and the intensity with which the neuron fires, and the overall shape of its response which it sends to the cells connected to it, also provide information to the brain (Figure 4). Even those neurons that do not change their activity when an odour is encountered, because the smell cannot bind with the receptor, contribute to the overall pattern of activity produced by each

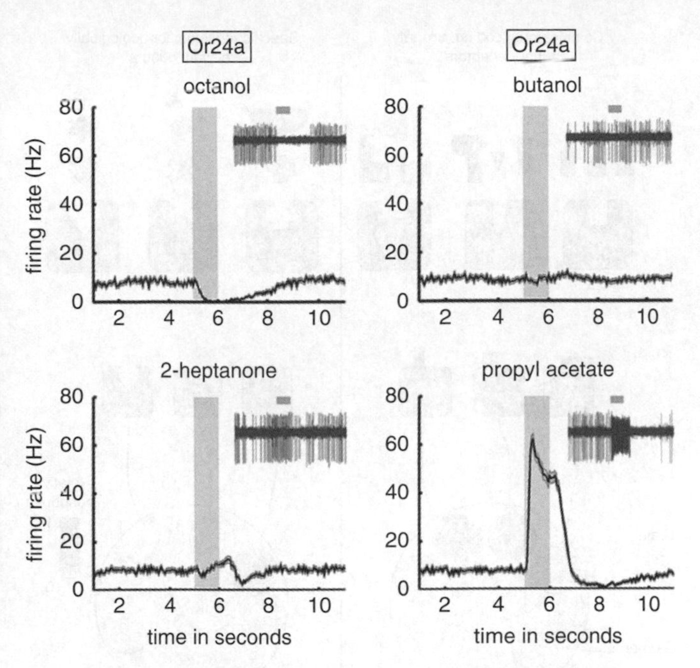

4. Varying responses of a particular *Drosophila* olfactory neuron to four different odours. The one-second odour delivery period is shown by a grey bar. The figures show the pooled responses of 110 different neurons. The inset panels show typical responses of a single neuron.

odour—no change in the activity of one class of olfactory neuron may provide significant information to the brain about what odour is being detected. This rich and complex response profile in the periphery shows how, with 400 types of receptor, the human olfactory system could theoretically detect a near-infinite range of odours.

The anatomy of smell detection

Despite the differences in the way animals detect smells—with noses, antennae, and various other organs—we all process the

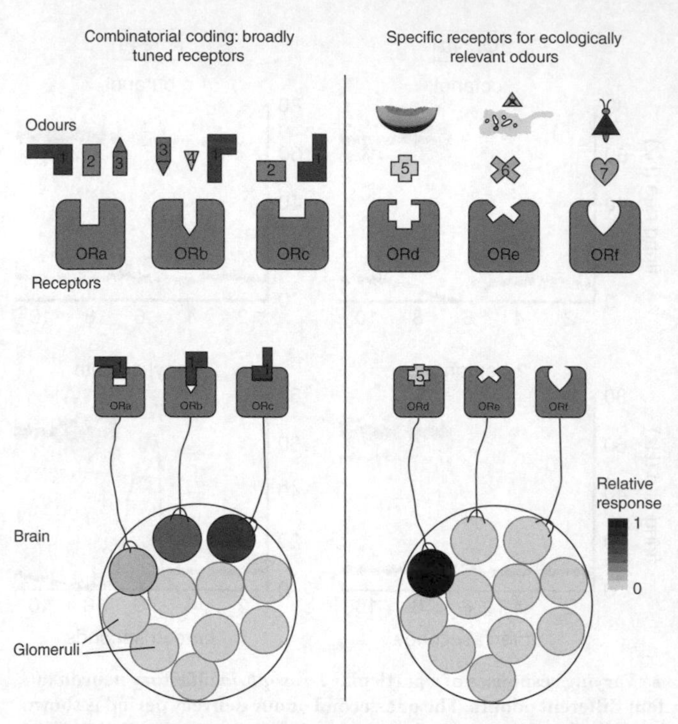

5. Schema showing the combinatorial code for general odours (left) and more specific responses to ecologically significant odours (right). Each of the general olfactory receptors (ORs a, b, and c) responds to more than one odour, with different degrees of intensity. Each of the specifically tuned receptors (ORs d, e, and f) responds only to a single, ecologically significant odour.

signals in a similar way. In some species, the distribution of different types of olfactory neuron across the surface of the olfactory epithelium, or the antenna, shows some degree of organization—one neuron type may be more densely distributed in some regions than in others. In other animals, such as fish, the distribution of olfactory neurons appears to be random. Where there is a spatial organization of these cells, it may mean that the brain has an idea of what kind of odour has been detected simply by being able to identify which region of the olfactory organ is responding.

Once it has been activated, each olfactory neuron sends its response into the brain, where the neuron converges with all the cells that express the same receptor type, to form a ball-like structure called a glomerulus. In a vertebrate brain there are hundreds or even thousands of these structures, depending on the species (Figure 6—the overall volume of the glomerulus is proportional to the number of neurons that project to it). In some animals there is one glomerulus per receptor type; in others, such as humans, there may be several glomeruli for each type. Furthermore, all brains in all animals are bilaterally symmetrical, so there are two sets of glomeruli, one on each side of the brain. This is a key difference between smell and taste in all animals—neurons that detect odours are directly connected to the brain, whereas those that detect tastes are not. Furthermore, the taste regions of the brain have no glomerular structure.

By gathering signals from many olfactory neurons of the same type, the glomeruli enable the animal to be more certain that a particular type of receptor has been activated. Just as the response of neurons is not binary (either on or off), the activity of glomeruli can be more or less intense, and also contains a temporal element—each glomerulus responds to different odours with activity that shows different intensities, time-courses, and durations. As a result, each odour/glomerulus combination produces a specific, unique, and repeatable output—the same glomerulus can respond in different ways to different odours, and the same odour can induce a different response in different glomeruli.

Where several glomeruli are activated by the same odour, those that respond first tend to be linked to neurons with the most sensitive receptors for that odour. These glomeruli, which respond within 1/10th of a second of the stimulus arriving at the receptor,

6. The two olfactory bulbs in the brain of the zebrafish, with the input from the olfactory nerve on the right. Within each bulb can be seen many glomeruli.

are those that seem to play the biggest role in enabling us to identify an odour. Some of the cells that take the signal from the glomeruli up into the higher areas of the brain respond to particular durations of glomerular activation—the difference between a whiff and a stench. In terrestrial vertebrates, the brain also has to be able to cope with changes in stimulation and sensitivity associated with the normal breathing cycle. Although sniffing can increase our detection of an odour, we do not normally perceive our sense of smell changing with every breath we take. Filtering out that uninformative breathing rhythm and providing a relatively constant perception is dealt with higher up in the brain. The same goes for smells that are constantly present—we eventually 'tune them out'. This highlights the fact that our olfactory perception, and its underlying mechanisms, have a high degree of plasticity.

Glomeruli located close to each other in the brain tend to respond to odours that have some similarities. For example, in the bee brain, glomeruli responding strongly to odours with the same functional group (alcohols, aldehydes, and so on) tend to be grouped together, with neighbouring glomeruli responding to chemicals with the same functional group but of different sizes. In other words, the spatial organization in the brain reflects aspects of the chemical structure of odours. There can be a functional aspect to this organization. In the mouse olfactory bulb, the fifty or so glomeruli that lead to an aversive response to the smells of spoiled food are grouped together at the top of the bulb, while those that appear to be involved in responses to food preferences are strung together in a long line going from top to bottom; glomeruli involved in pheromonal responses tend to be grouped together, with separate regions for those involved with reproduction, social behaviours, alarm signals, and so on.

Glomeruli interact with each other through neurons that gather signals from neighbouring glomeruli. These cells—called mitral cells and tufted cells in vertebrates, lateral neurons and projection neurons in insects—alter the activity and output of the glomerulus, depending on how neighbouring glomeruli have also been activated by the odour. Changes to the activity of glomeruli can involve what is called lateral inhibition—strong activation of one glomerulus may reduce the activity of a neighbour. This process, which occurs in many parts of the nervous system, sharpens the signal, increasing the certainty that a particular stimulus has been detected. The complex signal patterns that emerge from the glomerular layer of the brain are not simply a translation of the activity of the olfactory neurons; this processing in the brain is the first stage of our perception of smells.

One part of the output of the glomerular layer goes into brain regions associated with perception and the organization of appropriate behavioural responses; the other part goes to regions linked with encoding memory—the hippocampus in a mammal and the mushroom body in an insect. This means that the animal is continuously associating odours in the environment with things that are happening to it (Figure 7).

Exactly how the pattern of activation at these higher levels leads to the perception of a given odour is not clear in any animal. There are a number of theoretical models, revolving around the detection of synchronous activity in outputs of the glomerular layer—cells in higher brain regions will fire if they receive simultaneous input from a number of glomeruli. However, while this may enable higher regions of the brain to precisely identify odours, it does not seem to be involved in odour perception—these cells are activated up to 1 second after the first response of the olfactory neuron, whereas we can perceive a smell within one-fifth of a second of it being delivered to our nose.

There are occasional exceptions to this situation. A recent study from researchers in Israel surprisingly claimed that a number of women with apparently normal olfactory abilities turned out to have no detectable olfactory region in their brains when they were scanned. Either the relevant areas are in fact present, but are greatly reduced, below the level of resolution of the scanner, or this is an example of the great plasticity of the brain, with other areas of the brain able to take on an olfactory function, if the deficit occurs early in embryonic development.

It is striking that all animals with a brain share the same basic wiring diagram for detecting smells: OSNs, glomeruli, and higher neurons. This may be because this is the way it has always been

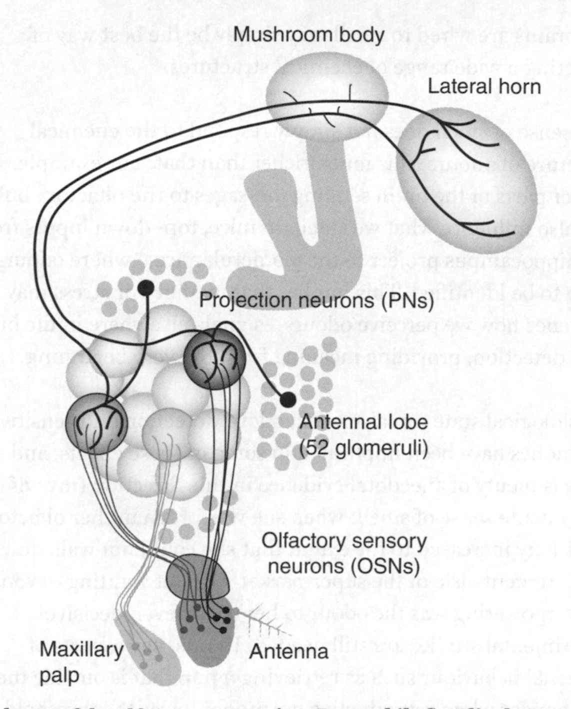

7. Schema of the olfactory system in a *Drosophila* fly. Flies have two olfactory appendages on their head—the antenna and the maxillary palp, which is found near the fly's 'mouth'. The antennal lobe is the part of the insect brain where the glomeruli are found. The mushroom body is the brain structure associated with memory, while the lateral horn organizes behavioural responses to odours.

done. The last common ancestor of you and me and a fly lived over 700 million years ago; if that early animal had a brain, it may have bequeathed the olfactory wiring diagram to us all. But it is perhaps more probable that the common ancestor of all animals did not have much of a brain worth talking about, and that the similarities in how our sense of smell works are more to do with what is called convergent evolution—in separate lineages, natural selection finds the same solution to the same problem. The way our brains are wired to smell may simply be the best way of detecting a wide range of chemical structures.

Our sense of smell does not simply respond to the chemical structure of odours. It is much richer than that. For example, higher parts of the brain sending messages to the olfactory bulb can also influence what we smell. In mice, top–down inputs from the hippocampus project to the glomerular area, where odours seem to be identified. This implies that memory or stress may influence how we perceive odours—signals elsewhere in the brain alter detection, providing meaning from the very beginning.

Physiological state can also affect odour detection and sensitivity. Hormones have been implicated in some of these effects, and there is plenty of anecdotal evidence in this direction (my wife has a very acute sense of smell; when she was pregnant her olfactory sensitivity increased to the extent that she could not walk down the detergent aisle of the supermarket without wanting to vomit, so overpowering was the odour to her). However, decisive experimental studies are still needed. In rats, key aspects of maternal behaviour such as retrieving a pup that is outside the nest appear when reproductive hormones lower the threshold at which the smell of a pup can be detected, thereby triggering the behaviour, which is absent in males or non-pregnant females.

Finally, smelling in the real world requires us to detect odours against a background of other smells and to respond to blends of odours. These are the kind of things that are hard to study in the laboratory, but which are fundamental to our everyday experience. We can train ourselves to discriminate different components in a blend—the reports of wine-tasters and the subtle aromas they say they can pick out are often the product of a great deal of training and attention. We can learn to smell.

Chapter 2
Smelling with genes

The beginning of our modern understanding of smell can be dated quite precisely. In April 1991, Linda Buck and Richard Axel at Columbia University in New York reported that they had identified genes in the rat that encoded proteins that they predicted were olfactory receptors. Although they had no direct proof, scientists all over the world soon showed that they were right. Buck and Axel's discovery transformed our understanding of the sense of smell, providing new tools for investigating how it works. In 2004, they won a Nobel Prize for their research. By focusing our attention on the genes that encode olfactory receptors, Buck and Axel opened up the possibility of understanding the deep evolution of olfaction, and of comparing smell across the whole animal kingdom—including in long-extinct organisms.

The receptors that Buck and Axel identified were part of the G-protein coupled receptor family of receptors that are widespread in our bodies. These proteins snake in and out of the cell membrane, with seven regions or domains sitting within the membrane, and the rest of the protein forming loops on the inside or the outside of the cell (Figure 8). Buck and Axel were looking for genes that produced proteins with this GPCR-like shape and that were expressed in the nose of the rat. It was already known that mice with a defective olfactory G-protein could not smell—this strongly suggested that vertebrate olfactory receptors were a kind of GPCR.

By using the DNA sequence of Buck and Axel's rat receptors, researchers were rapidly able to identify similar genes encoding olfactory receptors in a wide variety of vertebrates, including mice, frogs, and humans. But it soon became clear that there was something odd going on—attempts to isolate similar genes in insects, such as flies and bees, repeatedly met a dead end. My French colleague Renée Venard spent a fruitless year in a US laboratory trying to identify fly olfactory receptor genes using the data from Buck and Axel. She returned empty-handed, having inadvertently isolated only genes coding for olfactory receptors from mice (presumably due to a rodent infestation in the lab) and humans (probably from her own DNA).

These kinds of difficulties were a challenge to researchers; to get to the bottom of the enigma a race began to identify olfactory receptor genes in Drosophila, involving scientists in Axel's laboratory and a rival group in Yale led by John Carlson. In 1999, within weeks of each other, the two groups identified the genes that encode olfactory receptors in flies and got some insight as to why it had taken so long to make this breakthrough. Insect receptor molecules have a very different structure from the receptors found in the rat nose and turned out to have a separate evolutionary history. Although the proteins in vertebrates and those in insects are both called 'olfactory receptors', and both sets enable neurons to respond to odours, they work in very different ways. Insect olfactory receptors are not linked to a G-protein; instead, the receptor molecule directly allows the neuron to respond to an odour. In 2006, Richard Benton, working in Leslie Vosshall's laboratory at Rockefeller University, showed that although both insect and vertebrate olfactory receptor proteins have a similar transmembrane structure, they are oriented in opposite ways: the end of the protein that is inside the cell in a vertebrate is outside in an insect, and vice versa.

Families of olfactory receptors

Olfactory receptors in vertebrates and insects are not the only families of receptors that can detect smells—the known number of animal smell receptor families is continually growing, and we have surely not yet reached the end (Table 1). Each family of receptors tends to respond to a different set of odours. In vertebrates, for example, olfactory receptors are divided into two groups, class I (which detect soluble odours; these are the only type found in fish, but they are also present in small numbers in terrestrial vertebrates, including humans) and class II (which detect airborne odours and are found in terrestrial vertebrates; amphibians possess both classes). Many terrestrial vertebrates also have different kinds of receptors that are expressed in a special organ in the head called the vomeronasal organ, which processes many pheromonal signals. These vomeronasal receptors are divided into two families with different structures, known as V1Rs and V2Rs.

A bewildering array of other families of smell receptor have been identified in mice, some of which are also found in humans—trace amine-associated receptors (TAARs) which also use G-proteins, guanylyl cyclase-D cell receptors, MS4A receptors, and receptors found in a structure in the rodent nose called the Grueneberg ganglion. In mice, many of these receptor families appear to detect specific chemical signals that are closely linked to social behaviours.

The situation in insects has proved a bit more straightforward. There seem to be just two main families of receptor: olfactory receptors, which respond to a huge range of odours, and ionotropic receptors (IRs), which have a much simpler structure and detect compounds like ammonia. But confusingly, insects can also detect some volatile compounds using proteins that are normally classified as taste or gustatory receptors.

Table 1 Summary of known smell receptor types in vertebrates and insects

	Receptor type	Substance detected
Vertebrates	Olfactory receptors (ORs) Class 1	Water-borne odours
	Olfactory receptors (ORs) Class 2	Volatile odours
	Vomeronasal class 1 receptors (V1Rs)	Pheromones
	Vomeronasal class 2 receptors (V2Rs)	Pheromones
	Trace amine-associated receptors (TAARs)	Volatile amines
	Guanylyl cyclase-D cells	CO_2
	MS4A receptors	Fatty acids, steroids
	Grueneberg ganglion receptors	Pheromones/cues
Arthropods	Olfactory receptors (ORs)	Volatile odours
	Ionotropic receptors	Ammonia, CO_2 etc.
	Gustatory receptors	Cuticular hydrocarbons

Note: Not all animals in each group have all these receptors (for example, class 1 olfactory receptors are not found in terrestrial mammals, while arthropod olfactory receptors are found only in insects). The 'olfactory receptors' in vertebrates and insects are completely unrelated. Most of our knowledge comes from mice and *Drosophila*.

Understanding the structure of insect olfactory receptors has proved complex because the receptor molecule is composed of two components—a variable part, which enables the receptor to respond to a particular odour, and a protein known as Orco (olfactory receptor co-receptor), which provides the structural backbone to the receptor molecule. Orco is very highly conserved in insects—you can take a mutant fly that cannot smell because its Orco gene has been rendered non-functional, and then restore that olfactory function using an Orco gene from a moth (moths and flies separated about 250 million years ago). It appears that each insect odour receptor probably contains four protein molecules—three copies of the Orco protein and one copy of the variable receptor molecule. This is an extremely complex but active area of research, which raises the possibility of devising new strategies for combating insect pests and disease vectors by targeting the Orco molecule they all share.

How receptors work

We have a very clear idea of the smells some receptors respond to, because researchers have studied the responses of olfactory neurons directly—measuring their activity using tiny electrodes—or by studying the movement of calcium ions in and out of the cell (this is less precise). The odours a receptor protein responds to can also be studied by expressing the receptor either in cells grown in a petri dish or in a genetically modified hair on the antenna of the tiny fly *Drosophila*.

Exactly how olfactory receptors work—precisely how different smells are detected by different kinds of receptor—remains a mystery. The vast majority of scientists who work on the sense of smell assume that, just as Democritus argued, there is a link between the shape of a smell and how we perceive it. This is often called the 'lock and key' model, a metaphor that is not very exact because the key—the smell—can generally turn many different kinds of lock (or receptor), while each lock can be opened by more than one kind of key. Evidence in support of this model is seen in a rat olfactory receptor called I7, which is most strongly activated by molecules made up of eight carbon atoms but can also respond to molecules with seven or nine carbons. The receptor also responds differently depending on whether the smell molecule has one, two, or no double-bonds—these double-bonds change the shape of the molecule, enabling it to bind in different ways to the receptor.

Nevertheless, for the moment, we cannot look at the structure of a receptor molecule and predict what odours it will respond to. This gap in our knowledge, which will soon be filled, has led a handful of researchers to suggest that the lock and key model is wrong and that a completely different hypothesis, based on quantum mechanics and molecular vibration, is needed. Although media outlets can get excited about supposed quantum effects in the nose, none of the decisive experiments that have tested this idea have supported it—the most recent major experimental paper was entitled 'Implausibility of the vibrational theory of olfaction'. Ultimately, the issue will be resolved by crystallizing receptor molecules, studying their 3-D structure, and discovering how exactly they work. For technical reasons this has proved very difficult (this is true for non-olfactory GPCRs, too), but it will happen soon.

In natural situations, most olfactory neurons express just one receptor gene type, so each cell has just one kind of receptor on its membrane for detecting smells (there are some exceptions to this, for example in *Drosophila*). How neurons end up with only one type of receptor is still not clear, but in each cell it initially involves some sort of random selection of just one out of all the receptors that are in an individual's genome. A kind of negative feedback process then kicks in—the expression of the first type of receptor leads to the inhibition of the others in the same cell. Most animals have two copies of each gene, one from each parent. In your case, you have around 400 olfactory receptor genes, with two copies of each. It is possible that the versions of each gene you have inherited from your father and your mother differ very slightly; in that case each olfactory neuron in your nose somehow decided to express only one of potentially 800 alternative receptor types.

No one knows exactly how many copies of a receptor molecule there are on the membrane of a single olfactory neuron. In a talk given at a conference I co-organized in Manchester in 2011, the veteran insect electrophysiologist Karl-Ernst Kaissling estimated that there were perhaps 6,000 receptor molecules per square micron on the membrane surface of a moth olfactory neuron. If human receptors are present at a similar density, that would mean that each cell in your nose carries perhaps 6 million receptor molecules of the same type. You have around four million olfactory neurons, each one expressing only one of the 400 or so genes encoding olfactory receptors. If your olfactory receptor genes from each parent are slightly different, you have, roughly speaking, about 5,000 olfactory neurons of each type, and perhaps 30 billion receptor molecules of each type.

After the smell binds with the receptor, it is probable that either the smell molecule is rapidly removed from the receptor by enzymes that are floating about in the liquid medium that the neuron is bathed in, or that both the smell and the receptor protein are rapidly metabolized by the cell and a new replacement receptor is synthesized so that the cell can respond again. Whatever happens, it must occur quickly and at a very high rate—cells can respond very quickly to changes in odour presentation, and also carry on responding continuously to long-term presentation of odours. Even with millions of receptor molecules, and the probability that you only need a handful of receptors to be activated for the cell to respond, the olfactory neuron is a very busy place.

An example from humans

The significance of how a receptor molecule binds with an odour can be seen in our responses to an odour called androstenone. This smell is secreted in the saliva of male pigs and induces the sow to mate; it is also produced in human sweat and has led some unscrupulous manufacturers to market it as a human sex pheromone (it is no such thing). For many years, it has been known that people respond very differently to this odour—some people (like me) find it quite sweet, others think it is disgusting ('sweaty ball sack' is the most graphic description I have heard), while others find it quite sexy (this wide variation shows it is not a pheromone). People who do not like the smell of androstenone can be put off the taste of pork from uncastrated male pigs—this kind of meat is said to suffer from 'boar taint'.

In 2007, a group of researchers led by Andreas Keller and Leslie Vosshall of Rockefeller University showed that androstenone is detected by a single human olfactory receptor; in turn, this receptor is activated only by androstenone and a related compound, androstadienone. By sampling the DNA of over 300 subjects, the researchers were able to show that there was a link between how we perceive androstenone and the structure of the odour receptor protein, which is encoded by a single gene (Figure 8). Changes in two letters in the genetic code for the receptor are sufficient to alter the structure of the protein, and to change our perception of the odour from nice to nasty. By asking someone their response to androstenone we can predict their DNA sequence for this gene, and by looking at someone's DNA, you can predict their perception of the odour.

The structures of the slightly different versions of the receptor molecule present in different people must alter how the odour and the receptor interact and change the way that the neuron responds. In some mysterious way, that then gives rise to the very different perceptions reported by different people. Ultimately those differences in perception are nothing more than different forms of neuronal activity in the few thousand olfactory neurons that express this receptor.

Researchers have studied how changes in the structure of this receptor might lead to differences in the binding of androstenone to the receptor. All the changes that affect the response to androstenone occur in the part of the receptor that is predicted to be where the odour binds, exactly as the lock-and-key model suggests. Furthermore, using a chemical model of receptor function, researchers have been able to predict how particular mutations might affect perception of the intensity of androstenone. It is possible that this relatively simple odour–receptor link may

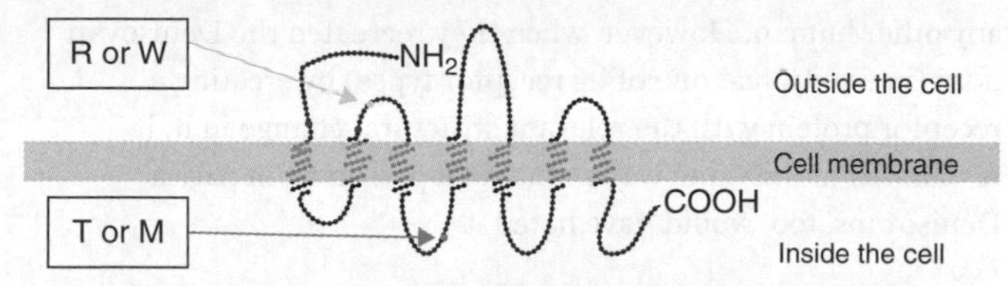

8. Schema of the structure of the human olfactory receptor that detects androstenone. Note the seven domains that are predicted to lie within the cell membrane. Proteins are made of amino acids strung together. The two most significant variable amino acid positions that affect our perception of androstenone are highlighted: on the outside of the cell you have either the arginine amino acid (R) or tryptophan (W), while on the inside you can have either threonine (T) or methionine (M). Both of these changes are caused by a change in a single letter of the genetic code. If both your copies of this receptor gene are of the RT form, you are more likely to find androstenone unpleasant.

provide an important clue to understanding how olfactory receptors function in general.

In one of the few really good scientific ideas I have ever had, I realized that we could use this link between gene and perception to explore how our extinct relatives, the Neanderthals and the Denisovans, would have perceived androstenone, by looking at their DNA. Ourselves, Neanderthals, and Denisovans all share the same genes—this enabled us to interbreed around 50,000 years ago—but in some cases we have slightly different versions of a given gene. My friend Kara Hoover (University of Alaska Fairbanks) and her colleague Hiroaki Matsunami (Duke University) took my idea and found that Neanderthals would probably have thought androstenone smelled disgusting—they shared the androstenone-detecting gene type that is particularly present in modern populations from the cradle of humanity, Africa, and which leads people to dislike the smell. Intriguingly, DNA from a Denisovan—a little-known group of extinct humans who lived in Asia and who were discovered only in 2011—had a completely novel version of this gene, which has not been seen in any other human. However, when they recreated the Denisovan nose (in reality, just one of its receptor types) by creating a receptor protein with the relevant structural change in it, it responded in the same way as that of a person from Africa. Denisovans, too, would have hated it.

This ability to peer back into the sensory world of our long-dead relatives is simply extraordinary, and very exciting. It will be hard to repeat the trick, because the link between receptor structure and perception is rarely so straightforward. By focusing on odours that are detected only by a handful of receptors—a recent study has identified a number of links between changes in receptor DNA sequence and alterations in perception of particular odours—it may be possible to gain more insight into how humans in the past would have responded to some smells.

How olfactory genes evolve

The similarities between the DNA sequences of olfactory receptor genes within a species can also reveal the speed of evolution—for example, in frogs and chickens there are sets of genes that appear to have evolved very rapidly, presumably responding to some change in the environment, such as the appearance of a new ecological niche. One of the features of smell receptor genes of all kinds is that they evolve very quickly. The underlying process is thought to be something like this: receptor genes are duplicated by mistake during the formation of egg or sperm; once there are two copies of a given smell receptor gene present in the genome, then random mutations can occur in the DNA sequence of one of those genes without affecting the animal's survival. By chance, some of those changes will alter receptor function slightly and enable the animal to detect new odours; if this increases the animal's ability to survive and reproduce, over the vast depths of evolutionary time this new DNA sequence will become fixed in the population. Where there was initially only one olfactory receptor gene there are now two, able to detect different odours.

A recent survey of olfactory receptor genes in fifty-eight species of mammals showed that gene duplication, followed by subsequent mutation, had played a significant role in the ability of herbivorous species to detect novel plant odours and to expand into new niches. Intriguingly, the researchers also found that solitary species such as sloths tended to have larger numbers of olfactory receptor genes. This effect is very variable: the largest number of olfactory receptor genes in any animal—over 2,000—is found in the highly social African elephant. What exactly all those receptors are doing is unknown—it is currently impossible to predict from the DNA sequence of a smell receptor what odours are detected by the protein produced by that gene.

Because gene duplication generally leads to the two copies of the gene being physically close to each other on the chromosome, the results of this process can be seen by studying the location of olfactory receptor genes in a genome. In humans, olfactory receptor genes tend to be clustered, reflecting past duplication events—chromosome 11 is a particular hotspot, hosting over 40 per cent of our olfactory receptor DNA sequences. Why smell receptor genes should apparently be so susceptible to gene duplication is unknown.

As in the case of the human gene that enables us to detect androstenone, many species show genetic variation for a given receptor gene, so different individuals within a population, and different populations within a species, have subtly different receptors, often leading to different responses to smells. For example, a study of wild *Drosophila* flies caught in Japan over a year—representing several generations—revealed that the frequencies of different variants of various olfactory receptor genes changed with the seasons. Another study investigated over seventy naturally occurring variations in a single *Drosophila* olfactory receptor gene sequence and found that many of these variants affected how the flies responded to odours. Taken together these data suggest that as food sources change during the year, different types of receptor are favoured, changing the genetic structure of the population and altering its attraction to smells.

The deep evolution of smell

Once the genes encoding the various families of smell receptor had been discovered, it became possible for researchers to look for them in the genomes of a wide range of animals. This has revealed, for example, that the family of olfactory receptors in your nose can be traced deep into the animal tree of life, as far as sea urchins, with which we last shared a common ancestor nearly 700 million years ago. Although virtually all of your olfactory receptors are Type II olfactory receptors—they detect volatile odours and appeared with the evolution of the first amphibians—the genes that encode these receptors are very closely related to those that produce Type I receptors which are found in fish (a small number of which you also possess), and to genes in sea urchins and starfish that look very much like our receptor genes. Something like these sea urchin genes may have been the origin of our sense of smell.

Genes involved in detecting smells also reveal the deep organization of the animal kingdom. Olfactory receptor genes in insects and mammals evolved separately to detect a range of odours; in the case of insects, they appeared after insects left the sea and primarily detect airborne odours. This can be seen by comparing insects with their parent group, the crustaceans (despite what you were taught in school, modern genetics shows us that insects are simply a weird kind of crustacean). Most crustaceans live in the sea, and they have no genes similar to the insect olfactory receptors—not even terrestrial crustaceans, such as woodlice. Similarly, millipedes, centipedes, and spiders, which left the sea before the insects, have no genes like the insect olfactory receptors. Instead, these arthropods have a large number of simpler ionotropic receptors for detecting volatile odours. These ionotropic olfactory receptors are in fact present in all bilaterally symmetrical animals except the vertebrates and the echinoderms (starfish and sea urchins, which are bilaterally symmetrical as larvae), which form a group called the deuterostomes (you are a deuterostome).

These analyses suggest that when the deuterostomes split off from the rest of the animal kingdom, over 550 million years ago, long before any multicellular organism was on the land, the only way animals had of detecting chemical information was probably through something like gustatory, or taste receptors. Sometime afterwards, the deuterostomes evolved Type I olfactory receptors for detecting dissolved odours, while the rest of the bilaterally symmetrical animals evolved ionotropic receptors. Eventually, after they had moved onto land, but before they could fly, the earliest insects evolved their olfactory receptors around 400 million years ago. These receptors are unique to insects—they are not even found in closely related arthropods such as springtails and bristletails, which make do with gustatory receptors and ionotropic receptors.

It is possible to combine genetics and comparative anatomy and come up with estimates for how many OR genes there may have been in long-extinct animals, and thereby gain some insight into the significance of smell in their ecology. In mammals, the size of the olfactory gene repertoire is highly positively correlated with the relative size of the cribriform plate—the perforated base of the skull through which your olfactory neurons dangle. From a fossil skull of the sabre-toothed tiger *Smilodon*, it is possible to predict the number of olfactory receptor genes it may have had (Figure 9).

Although this is impressive, the result is not entirely surprising—the figure of around 600 olfactory receptor genes is similar to that of a domestic cat. More challenging is to estimate how many olfactory receptor genes may have been present in a whole extinct lineage, such as the dinosaurs. Using the ratio of the size of the olfactory bulb to overall brain size and comparing this to the number of

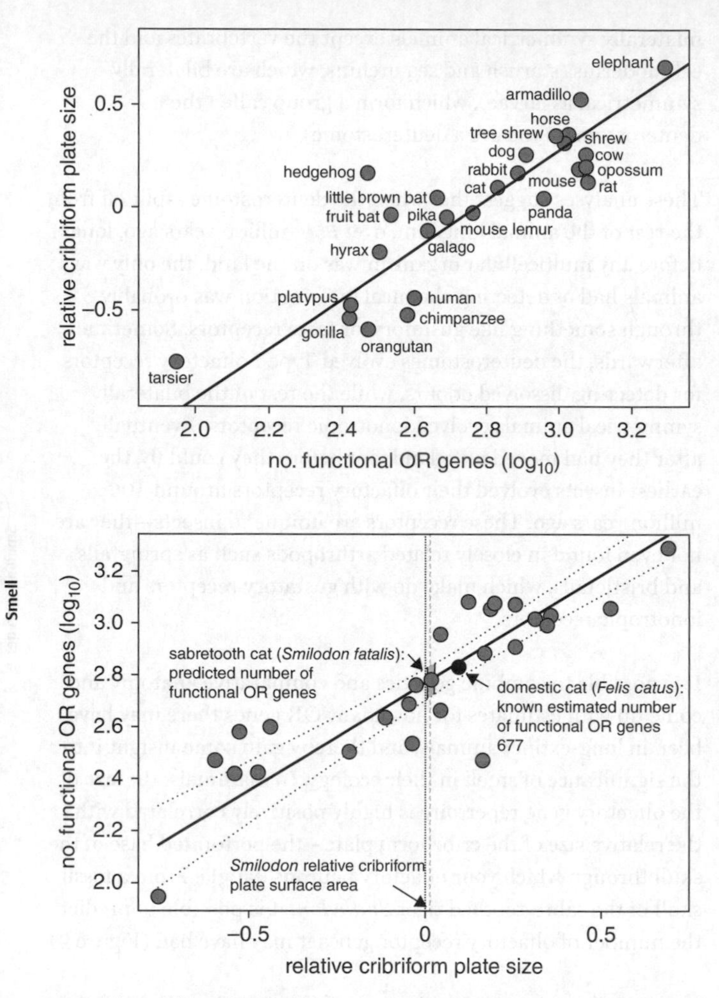

9. Number of functional olfactory receptor (OR) genes as a function of the relative size of the cribriform plate for a range of mammals (top) and position of the extinct sabretooth cat *Smilodon*, predicting it to have had around 600 functional OR genes (bottom).

olfactory receptor genes in a range of bird species, researchers have been able to estimate the number of such genes in extinct theropod dinosaurs (birds are just a lucky kind of dinosaur). The scientists found that theropod dinosaurs probably had at least 360–500 receptor genes, and that, for example, *Tyrannosaurus rex* probably had a larger number, with perhaps around 650—approximately the same number as a chicken, but about three times as many as a crow. In contrast, the modern alligator has over 1,000 olfactory receptor genes. We cannot tell how exactly *T. rex* used its large nose, but with so many receptor genes, smell must have been a significant part of its ecology.

Comparisons of smell receptor genomes can also reveal patterns of evolutionary history. The ancestors of mammals appear to have expanded the number of olfactory receptor genes at about the same time as they reduced the number of genes involved in vision, presumably as they became nocturnal to survive in a world dominated by dinosaurs. This tendency for nocturnal animals to have more olfactory receptor genes can be seen in modern animals—for example, the kakapo and the brown kiwi are both nocturnal birds from New Zealand and have many more genes devoted to the production of olfactory receptors than their day-living, or diurnal, relatives.

Exploring the smell ecology of an animal can reveal some surprising twists. The colugo, a nocturnal mammal from south-east Asia, has a far smaller olfactory gene repertoire even than many diurnal species. This is because although the colugo is active at night, its habit of gliding down amongst the trees in dense forest means that all its sensory focus is placed upon vision, rather than smell. This has resulted in a large number of what are called pseudogenes in the colugo olfactory genome. Pseudogenes are sequences of DNA that we can recognize as having once been a functional gene, but which now contain mutations that prevent a functional protein being produced. In other words, these pseudogenes are genetic fossils, remnants of some previous adaptation. In the mouse there are about six times as many functional olfactory receptor genes as there are pseudogenes; in the colugo, the situation is reversed, with about three times as many pseudogenes as there are functional olfactory genes. This indicates that natural selection is preserving olfactory genes in the mouse but is not doing anything to stop them gradually becoming functionless through random mutation in the case of the colugo.

Perhaps the most striking example of this loss of olfactory receptor genes can be seen in the cetaceans—whales and dolphins. The ancestors of these animals were terrestrial mammals, something like hippos; about 40 million years ago they began to return to the sea, and now they are entirely aquatic, although they remain air-breathers and therefore cannot smell under water. This means that the only time a whale or dolphin can smell is during those split seconds when it surfaces, exhales and then rapidly inhales. In cetaceans, around 68 per cent of DNA sequences that can be identified as olfactory receptor genes are pseudogenes. This suggests that olfaction plays a limited role in the survival of these animals, and that their olfactory genes can accumulate mutations without harming an animal's chances of reproduction. The percentage of olfactory receptor pseudogenes in other mammals is very different—37 per cent in the sea lion, which, although it spends much of its life in the sea, comes onto land to breed, 27 per cent in the dog, and 17 per cent in the cow. Similar effects have been observed in sea snakes, where species that give birth to live offspring in the sea have a far higher proportion of olfactory receptor pseudogenes than do species that lay eggs on land.

Sometimes the status of a stretch of DNA as a pseudogene is obvious—mutations have rendered the sequence meaningless because the genetic message is all scrambled. But more often the pseudogene is identified by the presence of a sequence of DNA bases that tells the cell to stop reading the gene (this is called a stop codon). Normally these stop codons are at the end of a gene; in a pseudogene they can be found anywhere, interrupting how the cell reads the gene. In 2016, Richard Benton and his colleagues at Lausanne discovered to everyone's surprise that, in some species of flies, mutations that produced a stop codon in the middle of a gene involved in detecting vinegar were simply ignored by the cell. The insect produced the receptor protein as normal, which functioned just as in other species where the gene does not have a misplaced stop codon. This was, as the title of the article put it, a 'pseudo pseudogene'. This phenomenon has been found in a number of other systems—including hints that it may occur in human olfactory cells, too—suggesting that we do not fully understand how cells interpret genetic instructions and raising questions about whether some OR pseudogenes are as inactive as their name suggests.

Olfaction in human evolution

The number of apparent olfactory pseudogenes in the human genome (54 per cent of olfactory receptor sequences) could be seen as evidence of a decline in the significance of smell in human ecology as compared to our ape cousins. In 2004, a group of researchers argued exactly this, and suggested that our ancestors began to lose functional olfactory genes at about the same time as they developed colour vision—they would have used vision, not smell, to detect good food sources. Our ape cousins, who do not have colour vision, have fewer olfactory pseudogenes, the researchers claimed.

This made for a good story, but it turned out not to be true. Within three years the scientists were publicly arguing among themselves over the technical detail of the paper, and in 2010 a separate group returned to the question and found no such effect. Using 551 olfactory receptor genes that were common to a set of primates, the researchers demonstrated that humans show no greater tendency to lose their olfactory genes than do chimps, orang-utans, or marmosets (Figure 10). Our genes tell us that our sense of smell is about what we would expect from our close relatives.

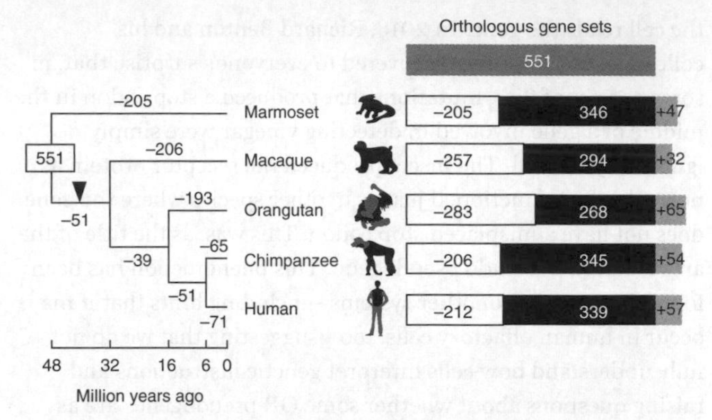

10. The number of olfactory receptor genes in five primates, based on 551 common ('orthologous') genes. At right are given the number of intact genes for each species, and with a minus sign the number of pseudogenes. Losses of functional genes at each step of the evolutionary tree are indicated by a minus sign. There is no greater tendency for humans to have lost olfactory receptor genes as compared to other parts of the primate lineage. New olfactory receptor genes have appeared in each lineage, as shown by the + numbers on the right.

As to what we have evolved to smell, some of our genes produce receptors that respond to natural oils and plant odours, as well as fruity and floral odours—similar genes are found in mice. The fact that these genes reach far into our evolutionary history implies a deep history of herbivory in mammals, which has long been known from anatomical evidence such as the structure of our teeth.

For the moment, revealing the full range of odours that can be detected by each of our 400-odd types of olfactory receptors (this process goes by the clumsy term 'de-orphanizing') is a far distant prospect, because there are so many odours that would have to be tested for each receptor. Identifying one or two odours per receptor type is much more likely and could probably be achieved within the next decade. Recently, a group of US researchers from several different institutions carried out the mammoth task of trying to find correlations between variations in olfactory gene sequences and behavioural responses to 276 different smell stimuli in 332 individuals. As expected, most of the olfactory receptors seemed to be linked to responses to many odours, but genetic variation accounted for 10–20 per cent of the perceptual variation in response to fifteen odours. In most cases, variation in perception related to perceived intensity, rather than the threshold at which an odour could be detected.

Because olfactory gene variations between individuals and populations may underlie differences in olfactory and flavour perception, they could also play a role in changes in diet over time and in different cultures. For example, people with the ancestral, African form of the gene that enables us to detect androstenone think that this substance smells foul. In 2015, Kara Hoover and her collaborators (including me) studied this gene in over 1,000 indigenous peoples from around the world and suggested that the mutations that lead people to think that androstenone smells sweet may have enabled the consumption of pork—people with this form of the receptor are less likely to complain of boar taint when eating the meat. One possibility is that the mutation arose in south-east Asia, which is also where we first domesticated pigs. Variations in our olfactory gene repertoire may help explain some key aspects of our ecology and culture.

Chapter 3
Smell signals

In the early 1670s, the pioneer Dutch entomologist Jan Swammerdam took a queen honeybee from one of his hives, attached her to a stick with a thread through one of her wings and watched as all the workers flew out to her. After a series of experiments like this, Swammerdam concluded that 'the female emits a very strong scent, by which the rest of the bees are attracted to her'. He had noticed similar effects in male silkmoths, which would start flapping their wings 'as soon as they get scent' of a female, in stallions that could smell a mare in heat, and in 'a parcel of dogs, which follow the female in her time of lust, excited by the bare scent'.

Although these examples showed that some animals use smell to attract other members of the same species, it would be nearly 300 years before scientists realized that these kinds of scents could usefully be grouped under a common heading. In 1959, Peter Karlson and Martin Lücher proposed a new term to describe these substances—pheromones. They explained:

> The name is derived from the Greek *pherein*, to transfer; *hormōn*, to excite. Pheromones are defined as substances which are secreted to the outside by an individual and received by a second individual of the same species, in which they release a specific reaction, for example, a definite behaviour or a developmental process.

Thousands of researchers now study pheromones in a wide variety of animals, mainly vertebrates and insects.

A few months after Karlson and Lücher published their proposal, their colleague Adolf Butenandt, a former Nazi who had won the 1939 Nobel Prize for his work on human sex hormones, announced the first chemical description of a pheromone. The stuff he isolated induced the wing flapping of the male silkmoth noted by Swammerdam three centuries earlier. After grinding up half a million female moths and testing various extracts, Butenandt showed that male silkworms were attracted to a sixteen-carbon alcohol. He called the substance bombykol (*Bombyx* is the Latin name for the silkmoth), and a single molecule of it can activate a neuron on the male's antenna.

Researchers soon realized that bombykol was not the only substance involved in the silkmoth pheromone—electrophysiological recordings revealed that only one of the two neurons in each of the tiny hairs on the male's antenna responded to bombykol. In 1978 it was shown that the other cell responded to a very similar sixteen-carbon compound, bombykal, which is present in the pheromone-producing glands of the female silkworm at a ratio of about 1:13 compared to the dominant bombykol. The silkmoth sex pheromone is primarily a blend of these two compounds, with species identity conveyed by the ratio of the components in the blend.

The effects of pheromones are generally divided into releaser effects (inducing an immediate behaviour in another individual) and primer effects (long-term alterations in physiology, and thence behaviour). However, in many cases the pheromone does not act in the kind of mechanical, stimulus–response fashion implied by Karlson and Lücher's definition, or by the term 'releaser'. Instead, these substances may be more accurately viewed as chemical cues which provide conspecifics with information about the identity or the status of a given individual.

As Tristram Wyatt has highlighted, such signature mixtures probably underlie many of the examples of mammalian chemical communication that we are used to seeing, for example the territory marking shown by cats and dogs. Although such signals do not conform to the strict definition of a pheromone—the responses they induce generally have an important learned component and may depend on the status of the receiver—they are clearly significant.

In the real world, the various aspects of how organisms signal to each other chemically may not be as clear-cut as neat definitions and striking examples suggest. The honeybee queen produces a cocktail of compounds that variously entice male bees to mate with her, attract workers, and also alter their reproductive physiology so that, while she is laying eggs and secreting her pheromones, they do not produce their own eggs. Like every other member of the hive, the queen is also covered in a slowly changing colony-specific signature mixture that is partly genetic and partly dependent upon the food the hive has consumed. This kind of complexity is probably the rule in most species—in only very few cases is there clear evidence that a single compound, say, attracts a mate and does nothing else.

Insect sex pheromones

The clearest examples of pheromonal communication have come from studies of insects. This is partly because so many species have been intensively studied due to their economic significance as pests, or, in the case of *Drosophila*, because it is a model for understanding basic biological phenomena. It may also be that the behavioural criteria for identifying pheromonal action—a clear response following stimulation—are easier to identify in insects. Compared to mammals they have relatively limited and inflexible behavioural repertoires that can be easily observed and linked to pheromonal stimulation.

Insect pheromones can be grouped into two main types—those that are highly volatile, and contact pheromones which require the individuals to come into very close proximity or actual contact to have their effects. Moth female sex pheromones, like bombykol and bombykal, attract male conspecifics over long distances. Sex pheromones have been identified in around 2,000 moth species, most of them variants on a similar chemical theme, constrained by the need for volatility. They can vary by length, functional group (acids, esters, alcohols, alkanes, alkenes), what is called saturation (the presence/absence, number and position of double bonds), and 3-D configuration. In general, they are 10–18 carbons long, with acetate, aldehyde, or alcohol groups, and between one and three double bonds.

Basic animal biochemistry means that similar compounds are used over and over again throughout the animal kingdom. For example, the female Asian elephant emits a sex-specific odour that also acts as a sex pheromone in over 140 species of moth. Obvious differences in the amounts of the molecule released by each individual explain why male elephants are not stomping around after female moths; female elephants are not surrounded by clouds of over-excited male moths because these species use pheromone blends—the elephant substance is just one component of the moths' blends.

The requirements for an insect to respond to a pheromone can be quite specific. In the tobacco hornworm moth, *Manduca sexta*, males will show a flight response only if they are stimulated with the right component molecules (there are eight in all, the main two being bombykol and bombykal), in the correct ratios and concentrations, and with an appropriate temporal structure. Continuous stimulation does not make the male take wing—in the real world such a strong dose of pheromone would indicate that he was right next to the female. In many species, the attractiveness of these sex pheromones is so strong that they can be used in agriculture to restrict pest levels. Males become so confused by synthetic female pheromone filling the air in and around the crop that they cannot find the real females, disrupting the likelihood of mating, reducing the number of eggs, and thereby limiting the number of destructive caterpillars.

A different kind of sex pheromone, detected by insects at very close range, was first discovered in houseflies and was then identified in *Drosophila* in the 1980s, through the activity of a French research group at Gif-sur-Yvette that I was a member of. This was the laboratory of my friend the late Jean-Marc Jallon, to whom this book is dedicated. These pheromones take the form of long hydrocarbons found on the fly's cuticle, hence their name—cuticular hydrocarbons (CHCs). They have since been found in virtually every arthropod in which they have been sought. These sticky, waxy substances change during the lifetime of the animal and help protect against desiccation. We initially assumed that CHCs evolved when arthropods first colonized the land around 400 million years ago, and that a role in sexual behaviour was gradually added to a protective function. However, these hydrocarbons have since been found to play a pheromonal role in marine crustaceans, which are closely related to insects, implying that the original function was in communication.

In *Drosophila melanogaster*, one fraction of female hydrocarbons induces male courtship behaviour—this precedes mating and enables potential sexual partners to identify each other and measure each other's quality. Jallon showed that when a male came into very close contact with a female, one particular set of hydrocarbons induced a key element of courtship, male wing vibration, during which the male fly produces a species-specific 'song'. However, subsequent research by my close friend Jean-François Ferveur (Jallon's one-time Ph.D. student) and myself revealed that things were much more complicated than we imagined.

Using genetic tricks, we made 'pheromone-free' females with no cuticular hydrocarbons and showed that the substances that we had spent over a decade studying accounted for only about one-third of male courtship. Another third of courtship is induced by unidentified compounds on the female cuticle, while the final third involves unknown volatile substances. The full chemical signature of a female fly involves several different kinds of stimuli, with hydrocarbons being only part of the picture.

Furthermore, to our great surprise, we found that although males of other *Drosophila* species did not normally court *D. melanogaster* females, when presented with our 'pheromone-free' *D. melanogaster* females, they became extremely excited and courted vigorously. We concluded that *D. melanogaster* sex pheromones excite their conspecifics, as you might expect, but they also inhibit inter-specific courtship. We also proposed that there is a common set of unknown, attractive pheromones that are possessed by all closely related *Drosophila* species. Researchers are now beginning to identify the precise neural circuits which enable each species to respond in an appropriate way. Peripheral detection seems to be identical; what differs is how the brain interprets these stimuli and how the fly responds.

Although many of the insect pheromones that have been identified are produced by females to excite males, sex is a two-way business, and males also produce pheromones. Male *Drosophila* are covered in cuticular hydrocarbons that are attractive to females and, in species like *D. melanogaster*, repel other males. In the case of moths and butterflies, volatile male pheromones are released by structures called coremata, which emerge out of the rear of the male's abdomen like a pair of tiny blown-up plastic gloves, and which diffuse his pheromone on the air. The pheromones produced by male butterflies often have a distinctive smell that we find pleasant (Table 2).

In some species, such as *Utetheisa* tiger moths, the female uses the precise levels of male pheromones to decide whether to mate with him. As caterpillars, these moths feed on plants containing toxic

Table 2. The scent of some male butterflies found in the UK

Latin name	Common name	Scent	Intensity
Lasiommata megera	Wall Brown	Chocolate cream	Slight
Melanargia galathea	Marbled White	Distinctly musky	Slight
Eumenis semele	Grayling	Sandalwood	Not strong
Maniola jurtina	Meadow Brown	Old cigar-box	Very slight
Arzynnis lathonia	Queen of Spain Fritillary	Heliotrope flowers	Slight
Lampides boeticus	Long-tailed Blue	Meadow-sweet	Slight
Aricia agestis	Brown Argus	Chocolate	Rather strong
Polyommatus icarus	Common Blue	Chocolate	Rather strong
Pierris brassicae	Large White	Orris root	Very slight
Pierris rapae	Small White	Sweet briar	Faint

Adapted from E. B. Ford, *Butterflies* (Collins, 1945).

alkaloids which they are able to tolerate, but which are repulsive to predators such as spiders and birds. Through the male's pheromones, the female can detect how much of these substances he has ingested; if the levels are high enough, she mates, and he transfers the substances to her, helping her and her offspring resist predation.

Detection

Volatile insect pheromones are generally detected by neurons on the antennae and are then processed in the brain. Male moths often have very large antennae, covered primarily with hairs containing two neurons, each tuned to one of the two primary components of the female pheromone blend. There are so many of these neurons compared to those devoted to detecting other odours that they form two immense glomeruli in the moth brain, known as the macro-glomerular complex. In *Manduca sexta* these two glomeruli sit atop the rest of the glomeruli—the one that responds to bombykal is shaped like a doughnut or toroid, the other, which responds to bombykol, is like a cloud or cumulus on top of the antennal lobe (Figure 11). The behavioural decision whether to respond to the output of these glomeruli is made higher up in the brain; in some species, higher neurons are sensitive to input both from pheromone glomeruli and from structures that process food odours, perhaps indicating that mating is more likely to take place when food sources are present.

Uncovering how cuticular hydrocarbon pheromones are detected has been more difficult. These heavy substances are sometimes not detected by olfactory receptors, so may not always strictly be 'smells', even if they can be detected at very short range. The receptor proteins that bind to these substances have not yet been identified, although some molecular components of the response to hydrocarbons have been identified in neurons on the fly's feet and proboscis. Potential receptors have even been found in neurons on the fly's wings, suggesting insects may be able to detect pheromones with many different parts of their body.

Complex pheromonal signals

The most intensively studied system of chemical communication is that found in *Drosophila*. The fly's world of sex pheromones is not simply composed of cuticular hydrocarbons—one good candidate for the common pheromone that Jean-François Ferveur and myself proposed must exist in many *Drosophila* species is a volatile substance called methyl laureate, which is detected by two types of olfactory receptor. Furthermore, the sex-specific *Drosophila* hydrocarbons that have been so intensively studied have recently been shown to be one step in a biosynthetic pathway that leads to highly volatile compounds which are also attractive to flies. All this shows that chemical signals are complex, and in

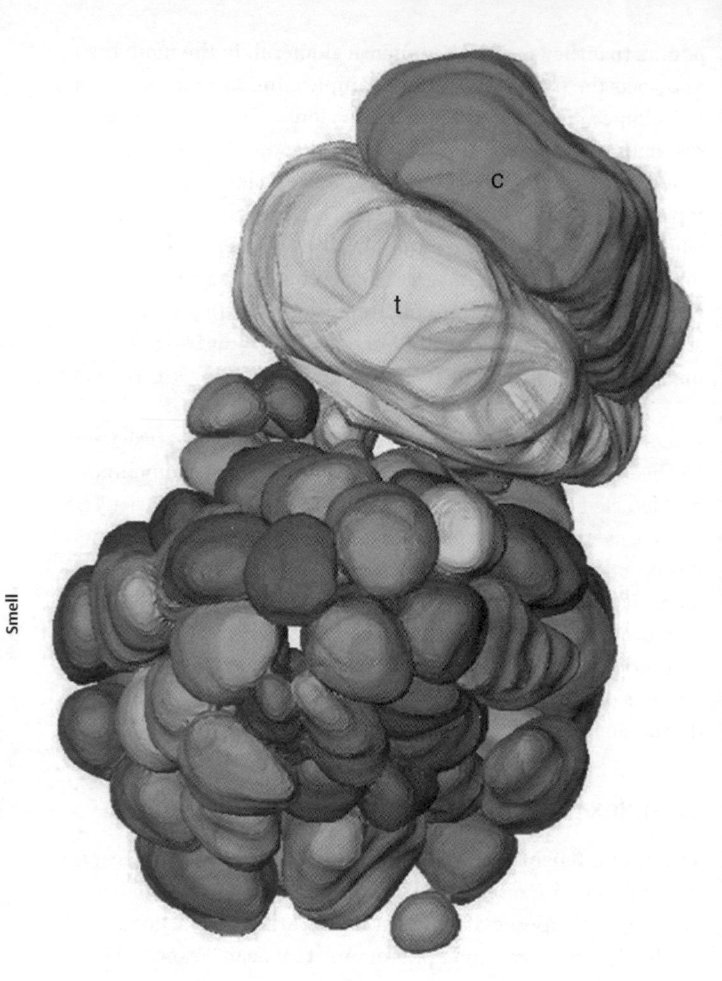

11. Computer reconstruction of the antennal lobe in the brain of the male moth *Manduca sexta*, showing the two macroglomeruli; the toroid (t) responds to one of the major female pheromone components (bombykal) while the cumulus (c) responds to bombykol. The remaining glomeruli can be seen below.

the same species may involve contact, short-range, and volatile pheromones. The same is true of those species that have primarily been studied in terms of volatile pheromones such as butterflies and moths—African squinting bush brown butterflies are covered with sex-specific hydrocarbons and courtship involves close contact as well as long-range volatile pheromones. Animals use a rich range of chemical communication systems to ensure that they are mating with the right individual.

Although identifying a sexual partner seems to be the main purpose of most pheromonal systems, on closer inspection things can turn out to be more complicated. In *Drosophila*, a sixteen-carbon volatile molecule called *cis*-vaccenyl acetate (cVA) has multiple roles. This compound is transferred from the male to the female during mating and was first found to reduce a female's attractiveness to other males after copulation (*Drosophila* males are generally not interested in mated females). As well as this 'anti-aphrodisiac' role, cVA is also used by the female, who introduces it into food when she lays her eggs, attracting other females to the egg-laying site. Although cVA is often called an aggregation pheromone, in fact it merely increases the attractiveness of food scents to fertilized females, through a small circuit in the female brain that is now well understood. *Drosophila* larvae are happiest when there are lots of other maggots on the rotting fruit they live on, churning it up and increasing the quantity of yeast, which is their food—attracting other *Drosophila* females to lay at the same site is therefore advantageous for everyone. This is not the case in species with carnivorous maggots: female houseflies introduce a repellent pheromone when they lay their eggs, putting off other females (and, presumably, other animals; my cats will not eat food that has fly eggs laid on it).

In *Drosophila*, cVA is detected by two types of neuron; the signals are processed in different ways in the two sexes, leading to different behaviours—the same substance is also involved in male–male aggression, which revolves around the male defending a particular patch of food against rival flies. Fly maggots can also detect cVA, as well as long-chain fatty acids and CHCs that form part of the same biosynthetic pathway. The natural history of *Drosophila* is not special, and we can assume that this degree of pheromonal complexity—which is still not fully understood—is mirrored in many other species.

For example, marine barnacles also have an interest in being close together—these crustaceans have a floating larval stage but settle down in adulthood, gluing themselves to the substrate. They spend the rest of their lives with their heads stuck to the floor, their legs wafting through the water to trap food. Most barnacles are hermaphrodite, so as long as they have another barnacle near them, they can mate, using the animal kingdom's longest penis relative to its possessor's body size. Because of their stationary adult lifestyle, barnacles therefore need to group together, and they are generally found gathered on rocks, on man-made structures, or on whales. The existence of a barnacle aggregation pheromone had long been suspected; it turns out that the glue that the barnacle uses to stick itself to the substrate is also the pheromonal 'settlement-inducing protein complex' that enables barnacles to gather together.

In social insects—bees, wasps, ants, and termites—a wide variety of pheromones help organize the society. In these insects, hydrocarbons identify sex and species, and contribute to the nest's chemical signature, thereby identifying nest-mates. In species with behavioural castes, these molecules can also identify what the insect does in the colony, mainly because both hydrocarbons and the tasks carried out by each individual change with age. These chemical cues are actually used by the ants—Deborah Gordon's group at Stanford University has shown that red harvester ants will not leave the nest unless they detect the hydrocarbon profile typical of patroller ants which normally guard the nest entrance, reassuring their sisters that all is well. If the patrollers are removed from the nest, the foragers will not leave. However, when Gordon's group rolled small glass beads covered with patroller hydrocarbons down the nest hole, the foragers emerged as normal (Figure 12). Many species of foraging ants leave pheromone trails on returning to the nest after a successful trip, enabling their nest-mates to find the food source—in the case of *Atta texana*, the trail pheromone is so powerful that one-third of a gram of it would leave a trace that could stretch around the globe.

Bees are pheromone factories, with multiple glands in their head, abdomen, and feet devoted to producing pheromones, each with a different form and function. For example, the venom gland, which is attached to the sting (this is what is pulled out of the bee when she stings you, killing her), not only produces venom but also an alarm pheromone made up of volatile short-chain esters that attract nest-mates to come and sting you. (If you are stung by a bee you should not grab the sting with your fingers and try to pull it out—you will almost certainly squeeze the venom gland, thereby injecting more painful venom into the wound and releasing even

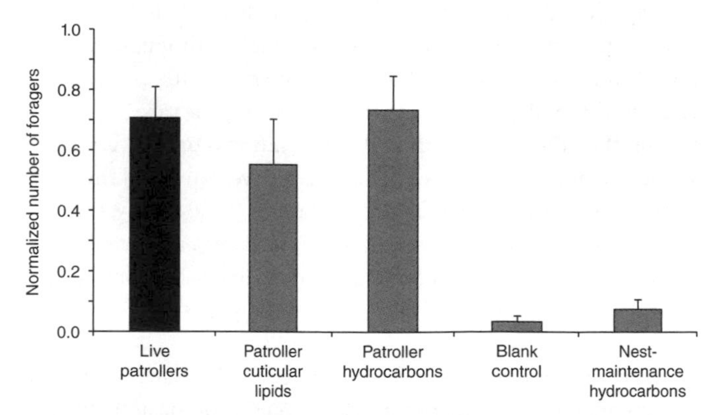

12. Number of foraging ants leaving the nest if presented with either live patrolling ants (far left column) or glass beads covered with various extracts. The cuticular hydrocarbons of patrolling ants produced the same effect as the actual ants.

more of the pheromone. Find some tweezers and remove the sting carefully.) During the honey bee waggle dance, in which a forager 'dances' inside the hive, using her movements to inform her fellow workers of the direction and distance of a food source, the dancing bee releases hydrocarbons that attract other workers to the dance floor, encouraging them to pay attention to her message.

In social bees, wasps, and ants, most members of the colony are female and virtually all of them are sterile, with non-functional ovaries. This odd situation, which worried Darwin, is generally thought to come about because all the members of the colony are very closely related. Bees, wasps, and ants have an unusual form of sex determination, such that if the queen has mated with only one male, her offspring—the workers—are more closely related to their sisters than they would be to their own offspring. They pass on more copies of their genes to the next generation by raising their sisters than they would do by mating and having their own babies.

In many social insects, if the queen dies or if she is removed from the colony by a callous scientist, previously sterile female workers will rapidly activate their ovaries and begin to lay their own unfertilized eggs which will develop into males. Although males do not work, they can leave the colony and mate with queens elsewhere, thereby ensuring that the workers' genes are transmitted. Pheromones underlie this shift in reproduction. Workers detect the presence of a reproductive individual through her chemical profile—depending on the size of the colony, this may involve a volatile pheromone (as in termites), a heavier compound like the honeybee's Queen Mandibular Pheromone, or cuticular hydrocarbons (as in some ants and wasps).

Scientists have puzzled over how to understand the action of these pheromones, which are produced as a function of the activity of the queen's ovaries. They may be a queen signal, providing workers with the information that a fertile queen is present, leading them not to reproduce in order to maximize the number of genes they pass on to the next generation by rearing their sisters. Or they may be a form of control, with the queen manipulating the workers' physiology against their genetic interests. Queen control would imply that there was potential for an arms race to take place between queens and workers—we would therefore expect some social insect species to have different ways of organizing reproduction, because the arms race would be at different points in different lineages with, in some cases, the workers having the upper hand. In reality, all social insects use the same way of regulating reproduction, suggesting that the queen control hypothesis is wrong. Further evidence in favour of the queen signal interpretation was found in 2014, when a group of Belgian researchers discovered that the same queen compounds lead to non-functional worker ovaries in a wide range of wasps and bees and ants. This deep evolutionary history reinforces the idea that these compounds are signals shared by many different lineages, not a manipulation by queens.

For pheromone evolution to occur, both stimulus and receptor must change simultaneously, requiring coordinated changes in the underlying genes. Genes encode proteins; although genetic changes may directly affect a receptor protein, pheromones are generally not proteins, so genes affect them indirectly, through enzymes that act on the biosynthetic pathway. In some elements of the *Drosophila* communication system the same gene encodes a receptor component and a biosynthetic enzyme, suggesting that in this example co-evolution of stimulus and receptor may have occurred relatively simply. This situation is not typical of most species, which use different genes to encode receptors and to produce the enzymes involved in pheromone biosynthesis. One hypothesis, proposed by Christer Löfstedt of Lund University in Sweden, is that response profiles of receptors in a given species may be broader than the range of substances produced, so that gradual change in a pheromone would not lead to the system breaking down. The receptor could still detect the altered form, while change in a broad receptor would not affect its ability to detect an unchanged pheromone. For the moment, pheromone evolution remains poorly understood, and there may be many ways in which this occurs.

Vertebrate pheromones

Identifying pheromones in vertebrates has generally been less successful, partly because their behavioural repertoires are more complex than those in insects, so a simple stimulus–response pheromone is less likely to be detectable. Nevertheless, there are many examples showing that vertebrates, too, use pheromones to influence behaviour.

In fish, many apparent pheromones belong to the class of what are called hormonal pheromones—these are modified forms of molecules that serve a vital role in internal reproductive physiology. The fish excretes them into the water, where they can be detected by conspecifics and play a role synchronizing the maturation of egg and sperm. This gives a clue as to how pheromones may have evolved long ago: if substances are released into the environment as a consequence of reproductive physiology, then any individual that can detect and act upon those cues may have an advantage, thereby increasing their fitness. Eventually, over evolutionary time, this can lead to a fully blown system of pheromonal communication (Figure 13).

In *Tilapia* fish from East Africa—and in trout, salmon, and goldfish—males excrete products in their urine that stimulate females, and which may exert a primer effect on the female reproductive system. In sea lampreys, which return to their freshwater spawning grounds after maturing in the ocean, a pheromone known as 3kPZS is produced first by young lampreys, attracting migrating adults, and then by males as a way of attracting females. The shifting role—first as a migratory cue and then a sex pheromone—has been demonstrated using synthetic compounds. The sea lamprey is a major invasive pest in the Great

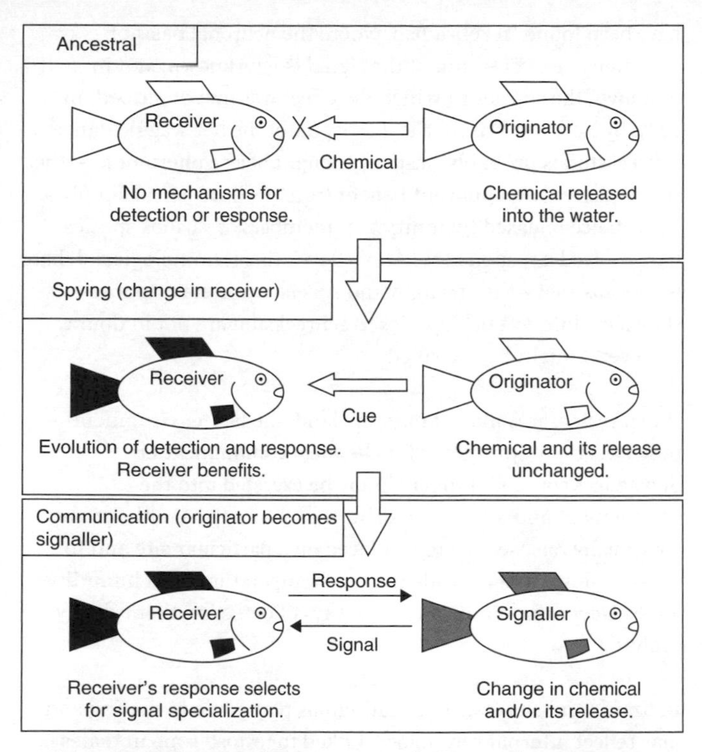

13. Hypothetical stages of pheromone evolution. In the ancestral stage (top), hormones and metabolites are released but not detected by conspecifics. In an intermediate 'spying' stage (middle), receivers have evolved the ability to detect hormonal compounds, but originators are not yet specialized for their release. In the final communication stage (bottom), both originator (now a signaller) and receiver are specialized for the exchange of chemical information.

Lakes, so there is a great deal of interest in understanding their system of chemical communication, in order to disrupt it.

Most dramatically, lampreys produce an alarm substance—a category known satisfyingly as schreckstoff—which causes intense agitation if the fish come into contact with it. Similar substances have been found in zebra fish, where the neuronal basis of detection and processing of the signal is now known, and in minnows, the species in which the effect was first described, in 1937, by Karl von Frisch, the discoverer of the bee waggle dance. Schreckstoff is probably best not thought of as a pheromone—the sender derives no apparent benefit from it—but instead simply as a substance released by injury that members of various species have evolved a response to. At one level, this terminological debate is very abstract—the strength and specificity of the responses shown by different fish species to schreckstoff are not in doubt, however it might be classified.

When vertebrates moved onto the land, the new environment posed a series of problems for chemical communication. Substances could no longer simply be excreted into the environment and allowed to drift; they now needed either to be continually released, or to be placed on a particular site and to slowly diffuse under a wide range of temperatures and humidity levels. New compounds and new ways of detecting them slowly evolved.

In lizards and snakes, chemical signals play a role in mating and may reflect internal physiology, linked to testosterone in males and oestrogen in females. Like many fish pheromones, and the queen pheromone in social insects, these substances have sometimes been interpreted as a chemical example of a special class of signals that ecologists call 'honest signals'—genuine reflections of physiology or fitness which cannot be faked or hidden. Whether this interpretation is correct will require more intensive investigation.

Reptile pheromones are often thick and evaporate slowly, having been rubbed onto rocks by individuals, thereby enabling conspecifics to identify their presence and perhaps their status. Pheromone detection in snakes and lizards generally involves the vomeronasal organ, which lies in the roof of the mouth. These animals gather odour particles using their forked tongues, which transport the molecules to the vomeronasal organ; neurons then project to the accessory olfactory bulb, a small area of the brain that has a similar glomerular structure to the adjoining main olfactory bulb. For many years it was assumed that pheromone detection in all terrestrial vertebrates relied upon the vomeronasal organ, implying that pheromones were absent in vertebrates without this structure, including crocodiles, birds, and humans. It is now known that vertebrate chemical signals can be processed by the main olfactory bulb—the absence of a vomeronasal organ does not necessarily mean that the species does not employ pheromones.

One of the most intensively studied reptile pheromones is a blend produced by female red-sided garter snakes under the control of oestrogen. This blend, which changes with age and physiological state, is very attractive to males. These North American snakes over-winter in dens and emerge in spring to mate, with males forming writhing mating balls around unmated females. Curiously, some male snakes may also produce the female pheromone blend and also become covered in writhing males. This poorly understood effect seems to be linked to temperature—as the courted males warm up, their attractiveness declines, perhaps because of changes in the volatility of their pheromone blend, and eventually disappears after a couple of days. If the male cools down, he once again becomes attractive. The exact function of this dramatic effect is still unclear.

Evidence for pheromones in other reptiles apart from snakes and lizards is less clear-cut, and no compounds have been identified. However, turtles, tortoises, and crocodiles possess glands on their heads or around their cloaca (the single external passage at the rear end of amphibians, reptiles, and birds) that produce various odorous chemicals; these animals also respond to water containing these products. Attempts to find proof of pheromonal communication in the tuatara from New Zealand have failed.

Most strikingly, despite weak claims about budgerigars, there is as yet no good evidence of pheromones in birds. Like crocodiles, birds do not possess a vomeronasal organ, so it seems very probable that their extinct dinosaur relatives did not do so, either. Whether that means non-avian dinosaurs did not use pheromones is unknown—there is no reason to imagine that the nose and the main olfactory bulb could not carry out this function. If a definitive answer can be obtained about birds—a single clear example of a pheromone would be significant, but proving a negative is difficult—then we would have a clearer idea what to think about chemical communication in the extinct dinosaurs.

Mammals

As Swammerdam noted, chemical communication plays an obvious role in the mating of domestic animals such as horses, cats, and dogs—males seem to be attracted to the odour of females in heat, and many mammals clearly use signature smells to identify themselves, or their territory. Surprisingly, we know very little about the actual substances involved, beyond that they are sometimes transported in urine. Detection of these signals often involves a striking behaviour called flehmen—the animal curls back its upper lip, inhales through its mouth and stares into space as though focusing on something (Figure 14). This action brings odours into contact with the vomeronasal organ, in the roof of the mouth. If you have a cat, you may have seen it doing this in response to the scent marks of other animals; male goats or horses near a female in season will also perform this behaviour.

The precise identity and function of pheromones in mammals remains poorly understood—for example, few of the chemical signals produced by these domestic animals have been identified. Olfaction scientist Dick Doty of the University of Pennsylvania has argued that if we adopt very strict criteria such as tests that employ synthetic versions of the putative pheromone, and the consistency of a behavioural response when stimulated, there is no

14. Stallion performing flehmen as he smells a female.

evidence that mammals use pheromones. Although Doty's insistence on experimental rigour is salutary, focusing too much on definitions may miss the point that chemical communication clearly occurs in mammals.

The most intensively studied mammal—the mouse—is often portrayed as a 'model' for understanding other mammals, including humans. As far as olfaction and pheromones are concerned, this is profoundly mistaken—the mouse is in reality very unusual. It has more than double the number of olfactory and pheromone receptor genes compared to the average mammal and has an unusual system for diffusing pheromones. In the area of chemical communication, it is, at best, a model for other rodents, and may possibly be just a model for other mice. Despite these limitations, research on mouse pheromones has revealed a number of important insights into the nature and detection of pheromones in mammals.

Mice produce a large variety of chemical signals, sometimes using pathways that were previously unsuspected. For example, a specialized set of neurons at the very front of the nose, known as the Grueneberg ganglion, detects both predator odour and a volatile alarm signal produced by other mice. If the nerve connecting this ganglion to the brain is cut, the mouse will happily explore a cage impregnated with the volatile alarm signal or with predator odour, whereas an intact mouse will freeze (Figure 15). In an indication of quite how complex chemical communication may be, some of the receptors on neurons in the Grueneberg ganglion have been suggested to be taste receptors that also detect volatile compounds. Mice also produce a variety of chemical signals that indicate their sex and social status. Female mice secrete substances that induce male investigative behaviour and mounting, while males produce a short protein called ESP1 that leads to female mating acceptance behaviour. ESP1 is detected by a common system in male and female mice, but each sex has a special brain pathway that leads to sex-specific behaviour.

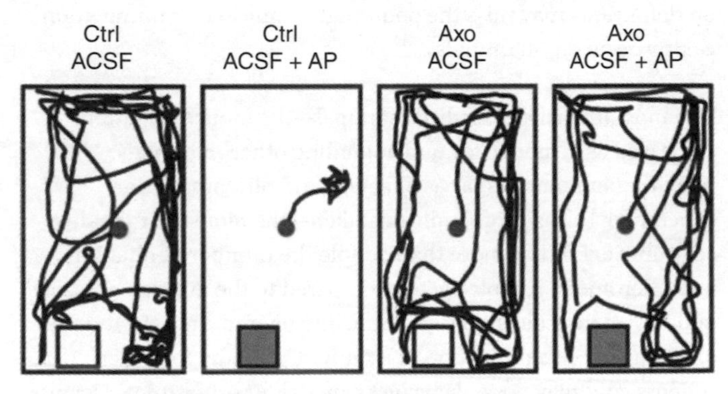

15. Effects of cutting the Grueneberg ganglion, a nerve in the mouse's nose, that responds to mouse alarm pheromone, shown as traces of mouse movement in a cage. Normal mice (Ctrl) will explore an open field (ACSF) but will freeze in the presence of the alarm pheromone (ACSF +AP). Operated mice (Axo) cannot detect the pheromone and show no change in their behaviour. The inset square signifies the source of the alarm pheromone; where the square is empty, no pheromone was present.

Male rodents produce thick, sticky urine that contains high levels of proteins, imaginatively called major urinary proteins (MUPs), which are rarely found in other mammals. As the urine evaporates, the MUPs remain, forming a crusty 'scent post' that diffuses odours revealing male identity and dominance. MUPs seem to act both as a way of slowly releasing small volatile organic molecules and to directly encode male identity. One particular MUP, called darcin (after Mr Darcy in *Pride and Prejudice*), binds and helps slowly release a compound called SBT that is found in male mouse urine and which stimulates female sniffing. This substance also acts as a male pheromone—a rare example of a protein playing this role in a mammal. The amount of darcin produced by a male reflects his dominance and seems to be used by other animals to identify who 'owns' a territory.

The study of mouse pheromones has revealed the error of the early assumption that the nose and the main olfactory bulb in the brain process 'normal' odours, while the vomeronasal organ and the accessory olfactory bulb process pheromones. There is a tendency for this to be the case—for example, male and female mouse odours are detected by a specific subset of receptors that are found only in the vomeronasal organ and are different from those found in the nose, and MUPs are detected by vomeronasal neurons. But there are also neurons in the nose that express TAARs, which seem to be involved in the detection of pheromone-like substances, and one potential pheromone found in male mouse urine, MTMT, is detected by the main olfactory bulb.

An example of the difficulty of working with mammalian chemical signals is shown by the Bruce Effect. In the early 1960s, Hilda Bruce noticed that if a pregnant female laboratory rat is housed with a male she has never met before, she will spontaneously abort her embryos and then mate with the new male. This effect involves low molecular weight chemical signals produced by intact mature males and has also been found in laboratory mice. Strikingly, it has not been observed in wild rodents, although there is anecdotal evidence for it occurring in a number of wild mammals, including gelada baboons. Above all, after more than half a century of study, and despite the undoubted role of olfaction, there is no clear explanation of which compounds are producing the effect, nor how.

Two examples show that other mammals apart from rodents do indeed use pheromones. First, male goats emit a blend of odours that have a primer effect, altering the female's reproductive physiology, in particular by the release of two key hormones, gonadotrophin releasing hormone and luteinizing hormone, which alter her behaviour and ovarian activity (Figure 16). Second, in rabbits, the mother spends most of her time eating, and

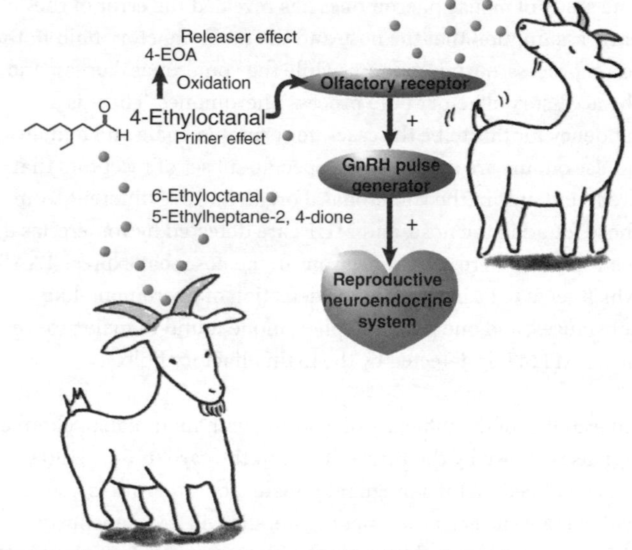

16. Schema of how male goat pheromones affect the reproductive physiology of the female, in particular the release of gonadotrophic releasing hormone (GnRH).

returns to the burrow for only a few minutes a day to feed her kits, which are able to find the nipple within seconds. In a series of careful experiments, Benoist Schaal of the University of Dijon showed that rabbit milk contains a volatile compound known as 2MB2, which is the nipple search pheromone and which is used by the kit to rapidly locate the nipple. If milk is deodorized it does not induce suckling, but if 2MB2 is then added to it, it regains its attraction. The response is also species-specific—hares, which spend the night with their leverets in an open-air nest called a form, do not need the kind of rapid access to nipples shown by rabbits and are not responsive to 2MB2.

As to humans, the vomeronasal organ, which has traditionally been seen as the site of vertebrate pheromone detection, is absent in adults (it is seen briefly during embryonic development but then disappears). Nor do we have the pheromone receptors found in mice. However, we do have some TAARs genes and there is no reason to think that we cannot detect pheromones without a vomeronasal organ. Much more significantly, there is no actual evidence that there are pheromones in humans. None of the studies that have claimed to prove the existence of human pheromones have either survived attempts to replicate them or led to a chemical identification.

The main substances that are frequently claimed to be human pheromones—androstenone and its linked compound androstadienone—induce no clear behavioural or physiological response and, as we have seen, there is substantial genetic variation in the population for our perception of both compounds. Attempts to link administration of androstenone with perceptual biases (describing a moving pattern of dots as more 'male' or 'female', for example) generally suffer from the same problems: the reported effects are very slight, they are based on a small sample of subjects, and they have rarely, if ever, been replicated. These problems can also be seen in other studies that claim to provide proof of pheromones in humans. For example, in 2011 research on twenty-four men suggested that administering the (imperceptible) odour of women's tears made the men feel less sexy and perceive a woman's face as less sexy. Three subsequent attempts to replicate the study have failed, and there is no reason to believe that the initial result was valid. The widespread belief that humans use smell to detect genetic similarity, and that we prefer the odours of people who are genetically dissimilar, is also based on weak and unreliable evidence.

One of the most famous studies of potential human pheromones was prompted by anecdotal evidence which suggests that young women who live together and are not taking oral contraceptives end up with their menstrual cycles synchronized. Over twenty years ago, a paper was published claiming that extracts of female armpit sweat, administered to twenty female subjects, produced mean changes in the length of the menstrual cycle of slightly less than two days. The statistical interpretation of these data has been challenged, and above all the study has not been replicated nor has the substance involved been identified. The anecdotal evidence seems to be based on humans being particularly attentive to coincidences and nothing more. Finally, attempts to identify a nipple search pheromone in babies have not produced clear results, partly because it is so difficult to separate out learning effects in humans—the babies in one study were aged between 65 and 86 hours, so would have had plenty of time to learn about any smell emitted by their mother's breasts, even those who were bottle-fed.

None of this proves that humans do not have pheromones, but it does indicate that scientists are often too hasty to claim they have found an interesting result when, in reality, there is nothing there. In the absence of any decisive evidence in support of the existence of human pheromones, the most interesting question is perhaps why we so readily accept weak claims about such a function in our species.

Chapter 4
Smell, location, and memory

One of the most famous stories about smell is that told by Marcel Proust in the opening pages of his sprawling series of novels, *A la recherche du temps perdu*. The narrator describes how, as a grown man, his mother makes him some tea; child-like, he takes a small piece of madeleine cake and soaks it in a teaspoon of the hot beverage. As he places the soggy cake in his mouth he is suddenly overwhelmed by a sense of the extraordinary, an exquisite feeling of happiness, the source of which he cannot immediately identify. Then he remembers—when he was a boy, his Aunt Léonie would give him madeleine dunked in tea; this in turn unlocks a whole series of complex and precise memories of his childhood. As Proust puts it, these unfold in his mind like Japanese paper flowers in a porcelain bowl full of water. I have had a similar experience, although, more prosaically, it involved the smell of hot Vimto and the vivid recollection of the café at Stockport baths.

The evidence that odours truly hold the key to complete recall is quite flimsy—the smell scientist Avery Gilbert has called it 'a literary conceit'. But, as many authors before Proust had noticed, and as scientific research has shown, smells can indeed release memories in a very powerful way. For example, there is good evidence that richer memories are evoked when adults are presented with childhood-related odours than with childhood-related images. The key aspect of memory that seems to be unlocked by smell—including in Proust's fictional example—is not simply a memory of a particular fact or a particular event, but of things or emotions that were experienced in a particular place and at a particular time. These memories do not have to be pleasant. In people with post-traumatic stress disorder, smells that are associated with trauma, such as napalm or blood, can evoke powerful fear-related memories.

The underlying basis of these kinds of effects is that in most animals, smells are used to immediately label experiences, so smell memories are often linked to places—to where a particular event occurred. In the mammalian hippocampus there are 'place cells' that are active when the animal is in a particular location and these provide a key to memory retrieval (the discovery of these cells was rewarded by the Nobel Prize in 2014). These cells are not simply a kind of GPS—they also integrate other sensory modalities, such as smell. Researchers have even created an olfactory virtual reality system for mice, revealing that place cells respond to an odour-guided virtual exploration of the world much as they do to a visual representation.

The way mouse odour memories are encoded depends on whether they are associated with a particular place, or with a particular moment in time. These separate 'when' and 'where' aspects of our smell-associated memories project to a brain structure called the anterior olfactory nucleus, which also receives input from the olfactory bulb and contains the 'what' aspect of sensory memory. This may explain why memories that are activated by smells can seem so vivid—in our minds we travel back to a particular place, often at a particular moment. Intriguingly, the anterior olfactory nucleus is a structure that shows accelerated degeneration in Alzheimer's disease, a condition in which people show both memory defects and a decline in their ability to identify smells. Many physicians are now interested in olfactory decline as an early indicator of Alzheimer's disease.

As the example of Alzheimer's suggests, in humans there is a strong link between memory, location, and smell. In 1953, Henry Molaison, better known to scientists as patient 'H.M.' (his identity was protected during his lifetime), had most of his hippocampus, amygdala, and entorhinal cortex on both sides removed in an operation that was intended to relieve his debilitating severe epilepsy. The result of the surgery was catastrophic—Henry could no longer create any new memories. Without his hippocampuses he could not recall anything that occurred following the day of the operation. For the rest of his life he lived in a perpetual now. His spatial memory was also severely affected, and he found it hard to read a map.

Strikingly, Henry performed normally when asked to say if there was an odour present, but he failed if he was asked to compare two smells, nor could he identify common foods based simply on their smell, even though he would have learned them long before the operation. So, for example, when presented with coconut aroma, he identified it as either soap or flowers, while the scents of mint, almond, and lemon were all described as flowers or 'an acid'. Even when he had a visual cue, things were still awry—once, when he sniffed a lemon, he said, 'Funny, it doesn't smell like a lemon.'

There is growing interest in the significance of the link between spatial memory and olfaction. A recent study of humans found that greater ability to identify odours was associated with better spatial memory, with frontal areas of the brain, which are involved in both olfactory processing and spatial learning, playing a particularly significant role. Patients with damage to these regions were less effective in identifying odours and in a spatial learning task, supporting the idea that olfactory identification and spatial memory may have common neural bases. Proust may have exaggerated the power of smell to evoke memories, but his suggestion that odours, time, and place are somehow connected in our memory was correct.

Navigation

These complex links between place, memory, and smell may be explained by an idea first put forward in 2012, by Lucia Jacobs of the University of California. She suggested that in all animals a primary function of olfaction is navigation. Jacobs's starting point was our difficulty in understanding why the size of the olfactory bulb in vertebrates does not always scale with the rest of the brain; associated structures, such as the hippocampus, also show this effect. The explanation may lie in ecology: a study of 146 species of terrestrial carnivorous mammal revealed that the relative size of the olfactory bulb is positively correlated with the species' home range size—the larger the area the animal normally covers in searching for food, the larger its olfactory bulb compared to the rest of its brain. Jacobs argued that the brain anatomy of different species with different foraging strategies also supported her hypothesis, and other researchers have adopted her framework in an attempt to understand the evolution of the vertebrate brain. The underlying explanation may be that the size of the olfactory bulb is directly related to the number of olfactory neurons, which in turn will relate to the ecology of the animal and the distance at which it detects odours.

Whatever the truth of Jacobs's hypothesis, olfaction is involved in animal navigation on both local and global scales. Pigeons can return to their loft even if they were released hundreds of kilometres away, and although the stars, visual landmarks, and even the Earth's magnetic field have been implicated in this ability, the sense of smell plays a fundamental role, in particular when the bird is only a few dozen kilometres from home. Scientists in Italy showed that pigeons with a damaged olfactory system were much less likely to return to the loft than those that were intact, while researchers in Germany mapped out the distribution of various odours around their laboratory in Würzburg, showing that what they called the olfactory landscape contained sufficient variation to account for the birds' homing ability.

Sea-faring birds such as shearwaters will return to their home burrow after foraging for days on end over the ocean. Researching the behaviour of such wild animals is challenging, but a number of studies have shown that birds with damaged olfactory systems cannot find their way home, and that most flights by three different species of shearwater involved olfactory-guided navigation (Figure 17). Members of the closely related species, the nocturnal blue petrel, which return to their burrow at night, also use smell to find their way home.

Fish such as salmon and lamprey will migrate across thousands of kilometres of ocean to find their way back to the stream where

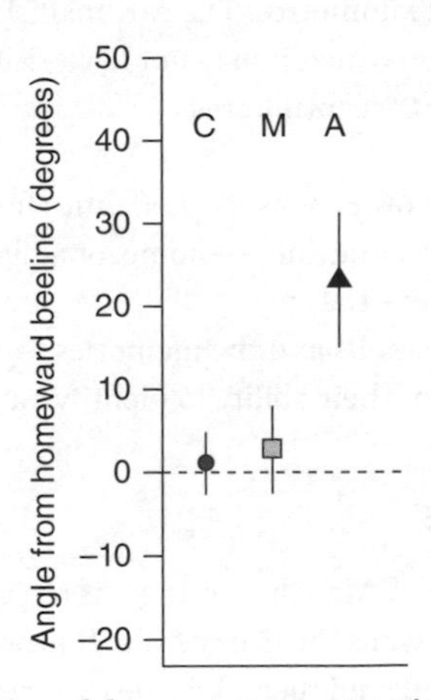

17. Homing orientation of shearwaters on returning from deep-sea foraging. An angle of 0° represents a straight line to the nest. C = control birds, with no manipulation; M = birds with a magnet on their head, to disrupt any use of magnetic information; A = anosmic birds with no olfactory system. Between 12 and 14 birds were studied in each group.

they hatched. As with migratory birds, it is probable that long-distance navigation involves other sensory modalities, but as the fish comes closer to the source of its home stream—perhaps even while out at sea—olfaction comes into play. A range of smells mark the location where the fry hatch, leading to rapid learning of this signal by the young fish. As salmon finally near their spawning ground, genes associated with detecting the odours of their home stream are expressed at lower levels, reflecting a decreased sensitivity to those odours, which are no longer needed to guide the animal to its spawning ground.

More localized homing in fish is also based on olfaction. The five-lined cardinalfish, a small tropical fish found throughout the Indo-Pacific region, feeds in open water at night, before returning to its home reef at daybreak, using olfaction to guide it back over distances of up to 2 kilometres. The cardinalfish also uses smell to find its breeding site, which it may not have visited in six months, returning to within 30 centimetres.

Humans clearly use other sensory modalities for most navigation, but when we return somewhere—home, or a place we have not visited for some time—the smell is both evocative and comforting. For many animals, smells and the memories they are entangled with are a key part of their ability to identify particular locations.

Locating smells

Near the University of Manchester there is a restaurant-crammed stretch of road known as the Curry Mile. It is not a mile long, nor is curry the only food sold there, but the experience of walking past the various cafés, restaurants, and eateries is an olfactory sensation. I find it impossible to go down the Curry Mile without continually sniffing, inhaling deeply to get as clear an impression as possible of the smells that are wafting by, moving my head from side to side to locate exactly which restaurant is the source of those enticing odours. Less pleasantly, I have used the same sniffing and moving technique to locate and destroy the source of cat urine when someone else's tomcat has invaded the house and left his mark.

These behaviours are exactly the same as those shown by many animals—raising the head, sniffing, and moving the head from side to side to locate the origin of a smell. In 2007, a study of University of California students showed that humans can track a scent trail of chocolate on grass by zig-zagging back and forth (if you no longer smell the odour, you turn back until you can). The students were blindfold, and moved along on all fours, with their nose close to the ground; the more frequently they sniffed, the quicker they completed the task. The experiment also showed that we can compare the signals from each nostril to identify where a smell is coming from.

Even a tiny *Drosophila* maggot can navigate using this smelling technique, moving its head from side to side, and using this information to locate an attractive odour. A maggot has small dome-like antennae, one on each side of its head (they are about 0.1 mm apart); activation of a single olfactory neuron in each antenna is sufficient for the animal to be able to move up an odour gradient. Matthieu Louis, then working in Leslie Vosshall's laboratory, showed that even though the differences in concentration arriving at each of the antennae would be minuscule, if a maggot has only one antenna, it is less effective at finding an odour source, because it lacks the ability to compare signals from either side. Insects like grasshoppers 'sniff' by waving their antennae in the air, while dogs track up an odour trace by zig-zagging from side to side through it, just like the California students. These lateral comparisons—in mammals and in insects—are made possible through the existence of neurons that compare signals across the two sides of the brain and also enable a unitary perception of the odour.

Experience and smell

Although the immediate structures that we use to detect smells—the receptor proteins on our olfactory sensory neurons—are genetically determined, that does not mean that everything about our sense of smell is hardwired.

Experience alters how we detect and process odours, altering the structure of our brains. For example, although in adult mice, all the olfactory neurons expressing a given olfactory receptor protein project to the same glomerulus in the olfactory bulb, in new-born animals a given type of neuron can project to several glomeruli. As the young animal smells, the excess projections are slowly pruned as some pathways are strengthened, and others decline. It seems probable that the same thing happens in humans, as our ability to process odours is sculpted by our experience. If animals are prevented from ever smelling, by genetically manipulating their receptors, the structure of the vertebrate olfactory bulb or, in an arthropod, the antennal lobe, is abnormal. Olfactory neurons need to be active in order to create a typical adult brain structure. This effect seems to involve the activity of proteins on the ends of the neurons, which attract and make connections to other cells, creating the glomerulus.

Olfactory neurons that are not stimulated tend to die. In mice, stimulation with a given odour leads to increased activation of the gene encoding the olfactory receptor that is expressed in that neuron; without stimulation these activation levels rapidly decline and the cells begin to die after a week or so. On the other hand, if stimulation with a given odour is continuous, over several days (this is clearly an unusual situation, but it might apply to odours such as those found in the mouse's nest litter), there is a decline in the number of neurons that detect that odour.

Experiments with rats show that their olfactory acuity—their ability to detect differences between two odours—can be improved by training. Single cells in the olfactory cortex, which is connected to the hippocampus, encode odour identity; in rats the activity of these cells was altered when an odour was coupled with a reward. Brain imaging studies of humans support this interpretation, suggesting that the olfactory cortex is actively involved in memory processing, and is not simply a site where odours are identified.

Such relatively simple experience-based effects can also be seen in our everyday experience, in particular when we stop responding to a continuous smell. For example, if you put some perfume on in the morning, you soon stop smelling it, even though other people will be able to smell it on you. The evolutionary explanation for this effect, generally known as habituation or adaptation, would appear to be that a continuous stimulus no longer requires attention to be paid to it—it can be safely ignored.

In the case of humans, the factors involved in the effect are complex, with molecular weight and the volatility of the molecule playing important roles, along with perceived intensity and pleasantness. Where adaptation with one odour produces a change in a response in another—this is known as cross-adaptation—we can assume that common pathways are being activated by each odour. In principle, this should provide some insight into the organization of olfactory processing, although exactly what that involves remains a matter of debate.

For many years it was unclear how adaptation works; my own assumption was that the key neurons involved were located at the periphery—olfactory neurons somehow became exhausted and ceased to respond to a continuous stimulus. I had a good-natured argument with Mani Ramaswami of Trinity College Dublin over this question—he insisted that, in the maggots we both study, adaptation occurs in the brain, as higher-level structures that collect information from glomeruli cease to respond. He was right. Work in my laboratory with Catherine McCrohan has shown that neurons subject to continuous odour stimulation will fire unabated for over twenty minutes. Even when a maggot was left in a high concentration of an odour for three days, and no longer showed a behavioural response to the odour, its olfactory neurons still responded normally. Other *Drosophila* researchers have recently suggested that, during adaptation, cells in and around the glomerulus alter their activity as a result of the unchanging activity of the olfactory neurons. In rats, the activity of mitral cells (the output of the glomerular layer) can change as a consequence of prolonged exposure to a particular odour, perhaps suggesting a similar mechanism is involved in adaptation in rodents.

One intriguing problem is how exactly we learn smells. In adult mice, olfactory learning seems to be based upon olfactory bulb neurons that were born in adulthood. In early life, the brain develops to process smells, creating structures that can identify and distinguish odours. It is not clear exactly how the new cells that appear in adulthood are involved in learning, and above all we do not understand how they are integrated into the existing brain networks.

The ability to learn to associate smells with particular events is widespread across the animal kingdom. In the 1970s, some of the early attempts by neuroscientists to understand the molecular bases of learning involved training tiny *Drosophila* flies to avoid attractive odours by pairing the nice smell with an electric shock. The molecules involved in olfactory memory formation in *Drosophila* turned out to be the same as those in other species, including ourselves.

The structure of olfactory memory

Memory is broadly divided into two types—short-term and long-term. One simple way of thinking about this is that short-term (or working) memory involves the electrochemical activity of a given neural network, whereas long-term memory involves changes to the structure of the network. In psychological terms, it is the difference between repeating a new phone number to yourself and just knowing a number you use often.

In *Drosophila* flies, different forms of olfactory memory can be identified. These range from short-term memory, which can be disrupted, for example by temporarily freezing the flies (a bit like turning your computer off without saving a file), through to long-term memory which can last for days and involves the rewiring of the fly's brain. In rats, despite the overlap between spatial and olfactory stimuli in memory, short-term olfactory and spatial memories appear to be encoded separately—rats can remember an olfactory memory while simultaneously being presented with a spatial memory task, and vice versa.

While human short-term memory has been intensively studied using images or sounds or numbers, there is remarkably little work on whether or not smells, too, are initially held in some kind of working memory buffer before being encoded structurally. Memory can also prime the brain to alter its responses to particular odours, by focusing attention on them. The general view is that we do have a short-term olfactory memory, although it is unclear how that then links with long-term memory—it may be that there is only one olfactory memory store which has different kinds of underlying physiological processes, corresponding to different memory time-scales.

In general, olfactory learning seems to be very rapid, unlike the situation with vision. For example, if people are presented with an image to learn, followed by several other images, the later stimuli interfere with recall of the target image. No such effect occurs with olfaction. Smells and flavours are involved in one of the strongest and least-investigated forms of learning—one-trial learning, where something happens just once, and is never forgotten. In an experimental context, if a rat is given some novel food, together with a drug that makes it sick, it will never touch that type of food again and it will avoid the smell. As a child,

I was sick after eating a meal that included cauliflower, and for the next thirty years or so the smell of the vegetable made me profoundly nauseous. In the case of rats, this learning can be indirect—cage-mates held with an animal that becomes sick after eating a novel food will also avoid that food, having smelled its odour on the animal's breath and sensed their cage-mate's malaise. Single-trial olfactory learning is not always so traumatic. Some ant species can learn to associate a neutral odour and a food source after a single presentation, with individuals remembering at least fourteen different such odours for the rest of their life. The evolutionary significance of this kind of one-trial learning is evident, but its exact mechanisms remain unknown, although recent work on *Drosophila* is beginning to unravel this widespread phenomenon.

A similarly strong form of learning occurs in animals such as sheep, where the mother rapidly learns the smell of its offspring and will generally reject attempts by any other lambs to suckle. This effect lies at the heart of the old shepherd's trick for pairing a ewe whose lamb has died with an orphaned lamb—the skin of the dead animal is placed around the orphan. With luck, the ewe will be duped into thinking that its lamb is still alive; eventually it will accept the orphan completely.

Human babies can identify smells that are associated with food their mother has eaten when they were in the uterine fluid, or, after birth, through the odour of breast milk. A study showed that new-born babies were more interested in the smells of anise if the mother consumed this flavour during pregnancy—the same effect has been observed with garlic and with the smell of alcohol. Odours learned in the uterus or at the breast are often associated with positive memories and can be recalled for years afterwards—they may well underlie the transmission of cultural food preferences and the strength of some of our attraction to particular foods.

Olfactory learning in insects can play a significant role in survival. Tropical leaf-cutter ants collect pieces of leaf, which they take back to the nest where they use it to grow fungus, which is the sole food source of the ant larvae. If the ants are fed food with a novel smell that contains a powerful fungicide, within a matter of days the ants have learned to avoid that smell. This effect seems to be mediated by odours emanating from the colony's waste dump—the ants quickly learn to avoid the novel odour, which they associate with the damaging effects of the fungicide. On the contrary, if a new odour is associated with a good food source, the ants will learn from the healthy smell of the fungus garden to be attracted to this odour. In the *Drosophila* brain, positive and negative associations are formed by separate pathways in the mushroom body, indicating a precise localization in the insect brain of different types of olfactory memory. All these examples underline that our olfactory systems are plastic, and change their activity depending on experience. Above all, they highlight the significance of olfactory learning in our everyday lives and in ecology.

Chapter 5
The ecology of smell

Our planet depends on smell. Because so many organisms can produce and detect smells, it is often a key channel through which species interactions occur, allowing for mutual benefit or exploitation and determining the shape of the world we live in. Some interactions are commonplace and simple, others are unusual and highly complex; all reveal the essential role that smell plays in making the ecosystem the way it is.

Pollination

Plants form the fundamental element of Earth's terrestrial ecosystem and over 300,000 plant species use insects to pollinate them, largely enabled by scent. The same is true of some types of fungi. For humans, many flower fragrances are attractive and give us pleasure. For pollinators, they indicate the source of a reward provided by the plant—primarily sugary nectar but also pollen itself, which is consumed by some insects. Attraction to the scent may be hardwired into the genes of the pollinator, but in some cases it is learned, as in bees. When insects learn to associate a new flower scent with a nectar reward, the way that odour is represented in the brain is altered—neurons taking the signal from the relevant glomeruli now fire at a higher rate. This suggests that, as in mice, there is some downward influence from higher structures in the brain that tells these neurons situated early on in the processing pathway that the signal now has increased significance. The way that even relatively simple brains represent smell at the earliest stages is not completely hardwired.

Although flower colour and pattern are involved in attracting insects, scent can be detected at long distance, out of direct sight of the flower, and in the dark. The powerful fragrance released by many flowers at dusk indicates that they are signalling to night-flying insects, particularly moths. Some bats also pollinate flowers as they search for pollen or nectar. Saussure's long-nosed bats are found in Mexico and use both olfaction and echolocation to find flowers, nuzzling for nectar in the flowers of succulents such as the iconic pipe organ cactus. In the few cases of plant–pollinator relations that we fully understand, flower fragrances are detected by olfactory neurons, as might be expected. However, even at the last minute, some insects use smell to be sure that they are heading to the right place—*Manduca sexta* moths have a neuron at the very end of their long proboscis, which they use to identify odours from deep within the flower.

Ecology is complicated and the ecology of smell is no exception. Any signal that is transmitted for detection by a receiver of the same species may also be detected by other organisms with more malign intents. Flower scent attracts pollinators, but it also reveals the location of the plant to herbivores, such as flower-eating species or insects that will lay their eggs so that their larvae can eat the plant. Sometimes, pollinator and predator can be the same—the adult takes the nectar, pollinates the plant, and lays her eggs on the leaves, which the caterpillars proceed to eat.

Plants can reduce their risk of revealing their presence to predators. For example, the smell of thistle flowers attracts both pollinators and flower herbivores such as the moth *Hadena bicruris*. As soon as a flower is pollinated, its bouquet, which is a blend of several different types of molecule, is subtly altered. Components that are particularly attractive to *H. bicruris* decline to near-zero, reducing the probability of attracting the attention of herbivores.

Because frequent insect visits deplete the flower's nectar and pollen reserves, rendering a visit fruitless for an insect, pollinators might be expected to pick up on any cues that could indicate that the flower had been visited recently. This appears to be the case in bumblebees, which inadvertently leave part of their sticky hydrocarbon profile on each flower they visit. The more bees that visit a particular flower, the greater the quantity of hydrocarbons that it carries. Older, more experienced bees can use these profiles to avoid flowers that have been frequently visited, the nectar stores of which are therefore more likely to be depleted. You have probably seen this effect in the summer: a bee will come very near to a flower and then move away without alighting. The fact that inexperienced bees do not avoid these hydrocarbon-covered flowers suggests that the older bees associate the hydrocarbon profiles on much-visited flowers with the disappointment of finding neither pollen nor nectar and have learned not to waste their time.

Some plants are dishonest and give insects neither a fragrant cue nor a nectar reward, but instead provide them with a simulacrum of sex. A number of orchid species do not produce a scent but manipulate male insects into pollinating them by imitating a conspecific female. This mimicry is partly visual—the plant grows structures that resemble a target female, sometimes to an astonishing degree—but above all it is chemical, with the plant producing hydrocarbons that correspond to key components of the female's hydrocarbon blend. *Ophrys sphegodes* orchids are pollinated by males of the solitary bee species *Andrena nigroaenea* who try to mate with the flower. The plant hydrocarbons are so similar to those found on the female that the male's antennal neurons respond in exactly the same way to the chemical profiles of both flower and female. Although the plant does not produce all the elements of the female's hydrocarbon blend, this presumably indicates that not all components are required for the male's strong behavioural response. Strikingly, the plant restricts the expression of these hydrocarbons to the flower—there are none on its leaves. The technical term for a deceptive odour like this, which benefits the sender and leads to the exploitation of the receiver, is an allomone. As we will see, the opposite situation occurs, too, where another species (often a predator) eavesdrops on a signal or cue and gains an advantage over the sender. These compounds are classed as kairomones. The terminology is much less important than the underlying process—deception in the case of an allomone, and eavesdropping in the case of a kairomone.

Some allomones are relatively simple. In the giant parasitic plant *Rafflesia* or in the titan arum (also known as the corpse flower, which, with its huge and highly stinky central florescence, is an attraction in botanical gardens the world over), components of the 'fragrance' resemble those produced by rotting meat. Flesh flies and other carrion-feeding insects are attracted to these odours, which include trimethylamine, found in rotting fish, and isovaleric acid, the smell of sweaty socks. The insects wander all over the flower, picking up pollen as they go; increasingly bemused at the lack of food or oviposition site, they eventually fly off, taking the pollen with them.

Other allomones are based on more complex interactions. In China, there is a white, daffodil-like flower that is pollinated by a local species of hornet. Researchers noted that the insect would pounce aggressively into the flower, inadvertently getting covered with pollen in the process, and then move onto another flower where the same thing would be repeated. Pouncing—and thence pollination—was induced by one particular substance in the flower's scent, which also induced high levels of activity in the hornet's antennal neurons. This substance is the alarm pheromone produced by the hornet's prey, the bee *Apis cerana*. In the normal state of things, the hornet eavesdrops on the bee's alarm pheromone and uses this signal to home in on its prey; the plant has evolved to manipulate the hornet's eavesdropping, and uses the same substance to attract the hornet, which gets neither food nor reward, merely unassuaged hunger and a dusting of pollen.

Chemical deception can also be used by plants to disperse their seeds. Female dung beetles collect mammal droppings and roll them away to a burying site, where they lay an egg on the nutrient-rich dung—when the larva hatches out, it has a ready-made food supply. *Ceratocaryum argenteum* is a rush-like plant found in South Africa which produces seeds that mimic the smell—and the sight—of antelope droppings, but which are too hard to be eaten by the dung beetle larva. The dung beetle wastes its time rolling a seed away from where it has dropped, increasing the plant's spread across the landscape, and then buries it. The seed sprouts, the baby beetle dies of starvation.

In each of these situations, the victims of deception can do little about it. Over thousands of generations of natural selection, the chemical mimicry of the allomone has become so precise that it is virtually impossible to distinguish between the original and the mimic. No longer responding to the mimic might avoid some wasted time and energy, but it would also mean not responding to the true signal of a vital food source or a potential mate, which would be far graver.

Predators and parasites

Prey location by predators is often relatively trivial—the prey either inadvertently emits an odour or it leaves a trace that the predator, be it carnivore or herbivore, can follow to its source. Plants typically produce volatiles that can be detected by herbivores—for example, Australian swamp wallabies are very sensitive to slight differences in the odour of the leaves of their preferred food, eucalyptus seedlings. Odour tracking is also common in many mammalian predators—even semi-aquatic mammals, like the star-nosed mole or the water shrew, use smells in underwater air bubbles to track their small invertebrate prey. Prey tracking occurs in invertebrates too—the larvae of *Pherbellia cinerella* flies, widely distributed in northern Europe, eat a variety of snails. The female fly lays her eggs on or near snail faeces and the larvae track down the snail, following the smell of its slimy trail.

In other cases, the predator takes advantage of the prey's olfactory preferences, in a process similar to that involved in deceptive pollination. Oil beetles are charismatic shiny black beetles found all over the world; their larvae, known as triungulins because they have three claws on each foot, parasitize the nests of solitary bees. The female beetle lays her eggs in a hole in the ground; when the triungulins emerge they move together up onto a twig where they gather in a small group. If they are lucky, they attract the attention of a passing male solitary bee; if he attempts to touch them, they jump onto him and he finds himself covered with grubs (Figure 18). When he mates with a female, the larvae transfer themselves onto her and are then taken by the mated female to the nest site she prepares for her offspring. Once she has laid her eggs, the triungulins jump off and begin eating their way through both the eggs and the food supply the bee has left for her babies; the beetle's well-fed larvae then pupate and eventually emerge as adults.

As you might have guessed, the beetle larvae mimic the cuticular hydrocarbon pheromones of the female bee, exciting the male and inducing him to try and mate with them. Instead he is sealing the fate of his offspring. In the case of a US oil beetle, *Meloe franciscanus*, male bees of the solitary species *Habropoda pallida* are attracted in flight to the hydrocarbons of the beetle larvae; the substances responsible for confusing the males are a blend of 23- and 25-carbon molecules with double bonds in position 9 or 11

18. Male solitary bee (*Colletes hederae*) covered in *Stenoria analis* parasitic triungulin beetle larvae. They will jump onto the female when he mates with her, then jump off her when she lays her eggs in a burrow and will eat them.

(Figure 19). The triungulins do not imitate all of the female's hydrocarbon profile, just those elements that are essential for the male.

In an equally tragic example, the South American bolas spider traps male moths by producing a long strand of web with a blob on the end that it swings round in the air. The web carries volatile hydrocarbons that resemble the pheromones produced by females of various moth species, attracting unfortunate males who think they will find a mate. Although the process seems pretty effective (the average spider attracts a male moth every six minutes), either the male moths are cannier than might appear, or the spiders find it tricky to reel in their prey—on average each spider catches only 2.2 moths per night.

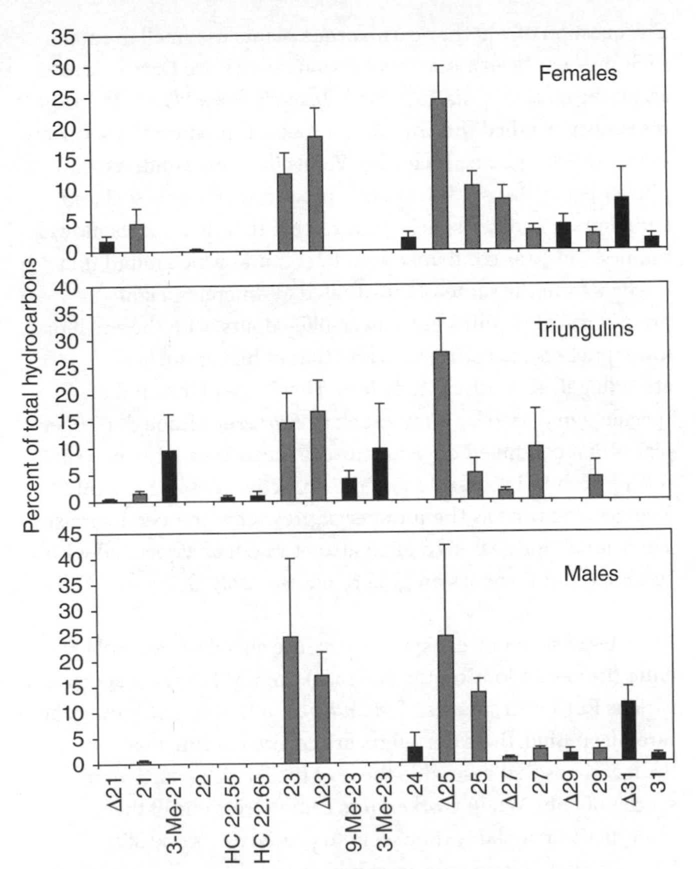

19. Cuticular hydrocarbons from male and female *Habropoda pallida* solitary bees, and from *Meloe franciscanus* beetle larvae ('triungulins'). For each hydrocarbon of a given length (21 carbons, 23 carbons, . . . 31 carbons), a number of variants ('position isomers') with double bonds on different carbon molecules are observed. Note the similar proportions of the main 23 and 25 carbon molecules in female bees and triungulins.

The question of whether carnivorous plants use smell to catch their prey has been a matter of dispute ever since Darwin wrote about the matter in his 1875 book *Insectivorous Plants*. In 2009, researchers studied the volatile compounds produced by a variety of carnivorous plants, including Venus fly-traps, sundews, and pitcher plants. Most of the plants produced odours associated with flowers in other species, suggesting that they may be flower mimics. This was confirmed by a later study, which found that pitchers from the same plant situated at different heights produced slightly different flower-like odours, with the smell of lower pitchers attracting ants and that of higher pitchers attracting flies. Neither study found evidence of the kind of specific signal used by plants such as *Rafflesia*, although pitcher plants that contained decaying insects unsurprisingly smelt of rotting flesh. It is possible that the attractiveness of the plant changes over time as the number of prey it has trapped increases. There is still no clear sign of an attractive odour associated with sundews, and Darwin's enigma remains unsolved.

Parasites of social insect species may use chemical camouflage to enter the nest, adopting the chemical profile of the host species. In various European species of *Maculinea* butterflies, including the rare Alcon Blue, the caterpillars are covered in cuticular hydrocarbons that resemble those of the larvae of particular species of ants. When worker ants come across one of these caterpillars on a plant, they seem to perceive it as one of their larvae, pick it up and transport it back to the colony where they feed it, even paying more attention to this insect cuckoo than they do to their own offspring. Sometimes there can be dozens of caterpillars in a nest—in small nests this can be quite damaging. Once it has emerged from its pupa, the butterfly pushes its way outside, mates, and the cycle begins again. There are hints that the ants might be on to this—one species of host, *Myrmica rubra*, shows geographical variation in its hydrocarbons, suggesting that some colonies may be less vulnerable to the very hungry caterpillar. A chemical arms race seems to have begun.

Sometimes, the chemical cuckoo takes over a whole nest of social insects. *Polistes* wasps are found in North America; they have small nests in which there is no queen, but instead a dominant worker who monopolizes reproduction. *P. dominulus* nests can be parasitized by females of a closely related species, *P. sulcifer*. The intruding female expels or kills the resident dominant female, and then takes over the colony. She does this by rapidly covering her body in the hydrocarbons that have stuck to the papery nest surface and within 90 minutes she is chemically camouflaged. The workers take her for one of their own and rear her offspring, even though they are not of the same species. An inverse process occurs in slave-making ants—these ants will raid the nests of other species, steal their pupae, and then bring them back to the colony. When the enslaved ants emerge from their cocoons, they act as though they are surrounded by kin and pursue their normal ant tasks, even though they are now working for a different species. This phenomenon can largely be explained by close or shared hydrocarbon profiles in the two species. This similarity in hydrocarbon profile probably explains the origin of slave-making—raiders attacked a nest of another ant species to find food, came across pupae that smelled like their own, and instinctively brought them back to the nest, thereby acquiring extra workers for the nest, workers that they did not rear. Free labour for the colony.

Parasitoids

There are hundreds of thousands of species of parasitoid insects, laying their eggs inside their living prey which is then eaten from the inside. So gruesome is this widespread lifestyle that Darwin famously wrote in 1860: 'I cannot persuade myself that a beneficent and omnipotent God would have designedly created parasitic wasps with the express intention of their feeding within the living bodies of Caterpillars.' Most parasitoid species are wasps, but flies, beetles, neuropterans, butterflies, and the weird strepsipterans (if you don't know what they are, Google them) also share this unseemly behaviour.

The main way the parasitoid finds its prey (generally a larva or a caterpillar, but sometimes an adult insect) is by odour, in particular through cuticular hydrocarbons. In the case of some Ichneumon wasps, this may involve using a long, slender ovipositor (perhaps twice as long as the wasp itself) that is bored into the trunk of a tree to find a beetle larva, which it seeks out using chemoreceptors on the ovipositor. The most fascinating uses of smell by parasitoids reveal more complex interactions. For example, when a butterfly lays its eggs on a leaf, this damages the leaf slightly and releases volatile organic compounds (VOCs). Parasitoid *Cotesia rubecula* wasps detect these VOCs in flight and orient towards them (in some species these odours are learned during the wasp's lifetime). *Cotesia* females prefer to land on leaves that carry eggs either just before or just after the eggs hatch—this is signalled by about a dozen compounds that are produced by the eggs.

In most cases, a wasp would gain nothing by laying an egg in a caterpillar that has already been attacked—the resident parasitoid larva would have a head start on eating the hapless caterpillar from the inside, leaving little for any late-comers, which might themselves turn into a meal. While *C. rubecula* larvae are in the host caterpillar, munching away, they leave a chemical mark in the host's spittle which female wasps can detect, and will avoid, preferring the smell of an unparasitized caterpillar. This mark is produced by the parasitoid's sting—its unique smell is produced by the venom and by a virus found in the wasp.

In some cases, a parasitoid wasp actively seeks out already-parasitized caterpillars—these species are hyperparasitoids, and their larvae will eat the parasitoid pupae inside the caterpillar. These wasps can detect the altered scent of the infested caterpillar's spittle—the complex smell ecology of their life cycle may therefore involve a plant, a host, a second host, and a virus. In species where parasitoid larvae do not mark the caterpillar, such as *Leptopilina* wasps, the female can actually measure how many eggs are in the caterpillar she is trying to lay her eggs in—the activity of the chemoreceptors on her ovipositor increases with the number of eggs, enabling her to avoid laying in prey that is already full of larval parasitoids.

Life finds a way

Just as predators can eavesdrop on prey, prey can eavesdrop on predators, altering their behaviour when they detect chemical cues produced by the predator, thereby gaining a benefit—in this situation, these cues are classed as kairomones. The Australian marbled gecko *Christinus marmoratus* eats less when in the presence of the odours of predators—both native (quoll, snake, dingo) and invasive (fox, cat)—presumably to reduce the probability that it will be attacked. Similarly, damselfly larvae (no mean predators themselves) are less likely to attack their prey successfully if they are in water that smells of a predatory fish. It appears that the larvae change their behaviour—perhaps they are less bold—in the presence of chemical signals denoting the presence of a predator. Bees avoid the smell of flowers where a crab spider—a camouflaged ambush predator—is hiding. They can detect spider hydrocarbons from up to a metre away, but sometimes the temptation to visit the flower is too great, and the spider gets its meal. These sub-lethal effects of predation, whereby stimuli released by a predator alter the behaviour of the prey, are increasingly studied by ecologists and involve the creation of what is dramatically but accurately called a smellscape of fear. These are regions where prey are more at risk of being attacked and which they therefore avoid, or in which they alter their behaviour appropriately, having detected the chemical cues of danger.

In *Drosophila* we now have a very precise idea of how these kinds of effects are produced. *Drosophila* larvae are attacked by parasitoid wasps, and will avoid the smell of wasps, while adult females will not lay their eggs when they detect wasp odour. The smell of wasp is detected by dedicated olfactory neurons and glomeruli—the function of one particular *Drosophila* receptor had been an enigma until it was found to respond to one of the key components of wasp odour. The activity of these olfactory neurons drives larval avoidance behaviour and in the female fly inhibits egg-laying.

Predator cues can trigger the development of defensive morphologies in their prey: in one species of the water-flea *Daphnia*—a tiny freshwater crustacean—this involves changing their morphology into a form that is so dramatically different that it was initially classified as a separate species. *Daphnia lumholtzi* can either show a typical form, with a round body, or it can grow a pointed 'helmet' and a long, sharp tail-spike. These inducible defences appear when the animals grow in water that contains fish predators. Although these defences are a response to predator-produced odours in the water, they are not simply a consequence of the crustaceans detecting their predator—the fish must have eaten *Daphnia* for the effect to work. Furthermore, the substances involved are not a kind of schreckstoff—putting ground-up *Daphnia* in the water has no effect. Instead, it seems that when the *Daphnia* are eaten, their crushed bodies mingle with the internal physiology of the fish—and perhaps its gut bacteria—to produce a cue that is released into the water by the fish when it excretes, and which other *Daphnia* can reliably interpret in terms of the presence of a predator. This phenomenon, known as predator labelling, has also been observed in tadpoles.

Similar kinds of anti-predator cues exist in plants. When the leaves of a plant are damaged, either by a browsing animal (large or small) or by egg-laying, organic compounds are released into the environment and can be detected by other plants, even members of other species. An experiment showed that if a sagebrush plant is clipped, nearby *Nicotiana* plants detect the relevant molecules and respond by producing more seed capsules, suggesting that these compounds can act as a common, cross-species alarm call, enabling other plants to prepare for a predatory onslaught. In the case of tomatoes, the signal becomes defence—plants infested by insect larvae produce an airborne substance that neighbouring plants absorb and convert into a chemical that restricts the growth of insect larvae. Plants can even use such odours to summon helpful predators—when a lima bean plant is infested with herbivorous mites it produces volatile compounds that are detected by neighbouring plants which produce more extra-floral nectar; this in turn attracts predatory mites which then attack the herbivores.

Not everything in ecology involves things eating other things. In commensalism, one organism benefits and the other either is not harmed, or benefits from the relationship. One of the most productive but mysterious of these relations has been the presence in humanity's dustbins of the tiny vinegar fly, *Drosophila melanogaster*, which was chosen by Thomas Hunt Morgan at the beginning of the 20th century for his pioneer work on genetics. It has since become one of the most intensively studied organisms on the planet. Although *D. melanogaster* now has a global spread, it originated in sub-Saharan Africa, where it coexists with a sister species, *D. simulans*, with which it does not interbreed thanks in part to the role of pheromones. One long-lasting enigma has been how, why, where, and when *D. melanogaster* became a human commensal—hovering around our bowls of fruit, teetering on a glass of alcohol, or breeding in our bins—whereas *D. simulans* did not.

To solve this problem, my friend Marcus Stensmyr of Lund University went on an expedition to an unspoilt area of Zimbabwe, near the probable origin of both fly species. The main *Drosophila* food sources in the area were two fruits—figs and a yellow fruit called marula. *D. melanogaster*, but not *D. simulans*, were highly attracted to marula fruit, and in particular to the main constituent of its aroma, ethyl isovalerate. Furthermore, *D. melanogaster* strains from the region have a variant of an odour receptor that renders them particularly attracted to this substance. Archaeological evidence shows that thousands of years ago, the San people who used to live in the area brought vast quantities of marula fruit into their caves, where it would ferment. Stensmyr's seductive hypothesis is that this would have attracted *D. melanogaster* flies to the alcohol and that over a long period these flies slowly adapted to their new environment and the dietary preferences of their close human neighbours. Thus began their long association with humans, and we subsequently took them all over the planet without even noticing. There appear to be two reasons why *D. simulans* flies, which also like alcohol, did not make that initial link with us: not only were they not attracted to the odour of marula, they also have an innate wariness of going into caves. In a field trapping experiment carried out by Stensmyr in Zimbabwe, *D. melanogaster* flies were all caught inside the caves, *D. simulans* were caught outside.

Humans as prey

For humans, the most dangerous animal in the world is the mosquito. Through the diseases they transmit, they inflict a greater mortality than any other single cause, including warfare, road traffic accidents, and pollution. Their lifestyle—biting humans—is relatively rare; there are only around 10,000 species of blood-feeding insect, and only about 100 species preferentially bite us. Although they are dangerous and annoying, mosquitoes are also an important part of many ecosystems, being eaten by a variety of other organisms as both adults and freshwater larvae. A key focus of many strategies to reduce disease transmission by mosquitoes involves understanding exactly how they track us down. This attraction involves smell—we all know of people who are bitten more than others, apparently because of the way they smell to the mosquitoes—but discovering the reality behind such

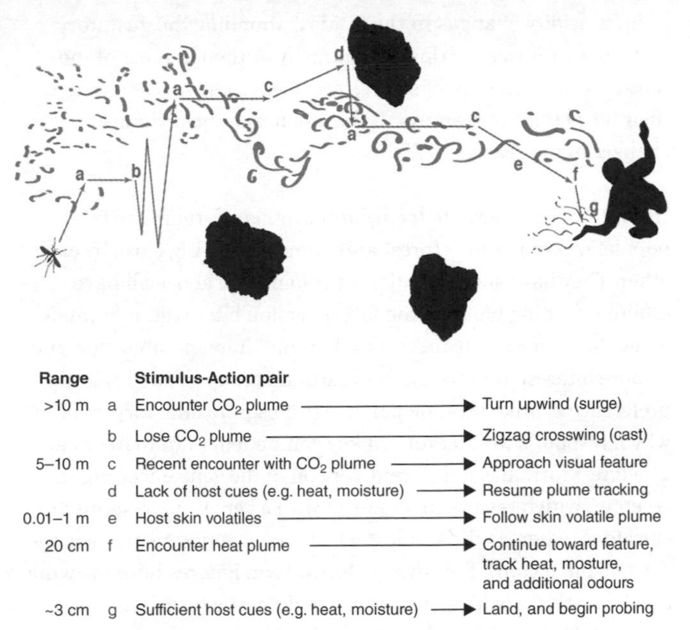

Range		Stimulus-Action pair	
>10 m	a	Encounter CO_2 plume	→ Turn upwind (surge)
	b	Lose CO_2 plume	→ Zigzag crosswing (cast)
5–10 m	c	Recent encounter with CO_2 plume	→ Approach visual feature
	d	Lack of host cues (e.g. heat, moisture)	→ Resume plume tracking
0.01–1 m	e	Host skin volatiles	→ Follow skin volatile plume
20 cm	f	Encounter heat plume	→ Continue toward feature, track heat, mosture, and additional odours
~3 cm	g	Sufficient host cues (e.g. heat, moisture)	→ Land, and begin probing

20. Summary of stimuli leading an *Aedes aegypti* mosquito (left) to locate a human (right), while navigating around vegetation (dark blobs).

anecdotes has been difficult and we still do not fully understand what precise compounds attract mosquitoes, nor why some of us are more attractive than others.

There are many species of mosquito that bite humans, each with their own ecology. Multiple senses are involved in how mosquitoes find us—olfaction is generally the most significant, but taste, vision, temperature, and humidity all play a role at different distances. Figure 20 summarizes the responses of *Aedes aegypti*, which transmits a range of diseases, including yellow fever, dengue, chikungunya, and Zika. Human odours are detected by mosquitoes using olfactory and ionotropic receptors and by gustatory receptors that detect CO_2, perhaps because the gas induces acidity changes in the fluid surrounding the gustatory receptors, or it may be detected directly. Although most of the mosquito's olfactory preferences are fixed, some species change their host target in response to host defensive behaviour or changes in the season.

In East Africa, where *Aedes aegypti* originated, there are two populations of the fly—forest and domestic—living close by each other. They have slightly different colours and above all have different feeding habits—the forest version bites wild mammals, while the domestic strain is found around human habitation and is more interested in us. Leslie Vosshall's group showed that this preference is linked to one particular mosquito olfactory receptor, which responds to sulcatone, a key component of human sweat. The forest form has a different version of the gene encoding this receptor, which results in a neuron with a very weak response to sulcatone, whereas the domestic form has a very sensitive version. This might account for why the forest form ignores humans while the domestic form loves us. Once the domestic form was established, these mosquitoes were inadvertently transported from Africa to the Americas, leading to repeated health crises, such as the Zika virus outbreak in Brazil in 2016.

Smell has also been at the heart of many attempts to reduce the biting activity of various insects. A widely used repellent, DEET, is generally quite effective although its exact mode of function is currently a matter of dispute—it may smell bad to the mosquito, it may confuse the insect's sense of smell by disrupting the olfactory receptors, or it may mask the odour it targets. It is possible that it works in different ways in different species. Further understanding how DEET works may lead to the development of even more effective ways of preventing insects from biting us.

Chapter 6
Smell in culture

Humanity's use of scents in culture can be traced deep into prehistory. Some of the most glorious images ever created were associated with fragrances—when people walked into the cave complex at what is now Lascaux, in France, around 17,000 years ago, they took lamps to guide their way and to enable them to create the stunning art that has miraculously been preserved on the cave walls (Figure 21). The wicks on those animal fat lamps were made of juniper and pine and would have given off a fragrant tang, mixing with the meaty aroma of the fat, the smell of the humans, perhaps sweating slightly from exertion and excitement, the musky whiff of their animal-skin clothes, the muddy odour of the paint, and the dry smell of the underground. Other twigs were available, but the people of Lascaux chose the fragrant ones.

The ancient tradition of burning scented woods, resins, or gums has a direct link to modern uses of fragrance—the word 'perfume' has its origins in the Latin words 'per' and 'fumum', meaning 'by smoke'. In antiquity the production of incense and the extraction of resins became a major industry—the ancient Egyptians burned different kinds of incense three times a day as an offering to the sun (resin in the morning, myrrh at noon, and a blend in the evening). Egyptian trade routes brought grasses from Syria, juniper from Phoenicia, and frankincense and myrrh from Somalia. These last two substances are both yellowish resins; the

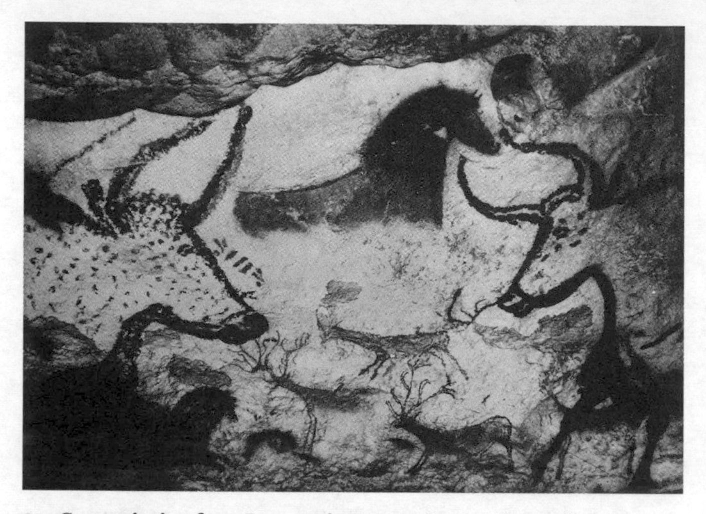

21. Cave painting from Lascaux in France. Among the smells associated with the creation of these images would have been the dry smell of the cave, the smell of the people in the cave, perhaps sweating slightly with effort or excitement and probably wearing pungent animal skins against the cold underground, burned wood (the darker pigments), mud (the ground-up ochre mixed with water), sizzling animal fat (the lamps that lit the work), and pine resin (the wicks chosen for use in the lamps).

smell of frankincense has been described as a mixture of pine and lemon, while myrrh smells vaguely vanilla-like. All these products were used not only in incense, but also in highly prized perfumes which, during the Roman era, became literally worth their weight in gold—hence the Bible story about the gifts of the Three Wise Men.

Beliefs in supernatural beings and locations also involved smells. The ancient Egyptians believed that the gods sweated incense; by covering a corpse in sweet-smelling unguent, the deceased would become closer to the gods. For the Greeks, the gods on Mount Olympus smelled of ambrosia and nectar, which they both ate and anointed themselves with. Christian writers generally described

Hell as being full of foul smells, while according to Buddhist belief evil flesh-eating demons give off a rank odour. In a number of contemporary cultures evil spirits are still described as emitting a stench, and in many cultures the dead body is anointed with perfume to attract angels and repel devils.

After Christianity became fused with the Roman state in the 4th century, incense—previously condemned as pagan—played a growing part in Christian beliefs and rituals. It was widely believed that priests smelled of a sweet fragrance that revealed their closeness to God, while saints exuded 'the odour of sanctity'. The 5th-century Christian saint Simeon Stylites, who famously lived on a pillar near Aleppo, allegedly produced a sweet smell that lingered around the base of his stone column, while when St Patrick died a lovely smell was said to have filled the room.

Much later, and on the other side of the planet, incense was used by the Aztecs for religious and cultural ceremonies; striking pottery incense burners or censers, often depicting mythological figures, have been found in Mayan and Zapotec sites (Figure 22). Incense was (and is) a key part of Buddhist rituals, with the Buddha often associated with the smell of sandalwood, while in Hindu traditions the goddess Lakshmi lives in the sandalwood tree. Spreading from China to Japan about 1,300 years ago, the use of incense became a key part of Japanese society, moving from the court into general culture, such that, eventually, different fragrances were used at different times of the day and in different parts of the home, a practice known as *soradakimono*. In Japan, smelling incense is an art-form known as *kōdō*, and is of similar cultural significance to flower arranging and the tea ceremony. It has also been turned into a number of games, such as *kumikō* and *genjikō*, in which players smell various kinds of incense and have to name the fragrance. In some cultures, incense is used as a form of social ceremony—in the 19th century, an English traveller to what is now Saudi Arabia described how, at the end of a meal or after drinking coffee, a box of incense would be passed round,

22. Drawing of early Mayan censer for burning incense. This object was found in Lake Peten Itza in Guatemala.

with each person inhaling the sweet odour and men opening their shirts to allow the fragrance to penetrate their body.

The smell of perfume

Using fragrances to scent the body is an ancient tradition, found in virtually all cultures across the world. Darius III, King of Persia in the 4th century BCE, had fourteen perfumers in his court; at the time, perfumes were worn in the hair, on the body, and on the feet—this tradition persisted, and can be seen in the story of Jesus's visit to the house of Simon the leper, where he had his feet perfumed with nard, a fragrant and very expensive ointment. The attractiveness of different kinds of perfumes has changed over time and place. In the Middle East and Europe, the first recorded perfumes were based on tree resins and gums; then the trend moved toward flowers, such as iris scents from Corinth, or marjoram from Cos. Perfumes based on animal products, such as the glands of musk deer, civet, and beavers, became fashionable in the late Middle Ages.

Much of the modern significance of perfumes—and indeed how they are produced—can be traced to the court of the French king Louis XIV, where they formed a central part of aristocratic displays of wealth and power. Official licences were granted to perfumers to produce fragrances under royal approval, spreading the link between prestige and perfume to those who were actually producing them. Perfumes were created by skilled artisans, who used their heightened olfactory senses and imagination to elaborate new and enticing fragrances. In the second half of the 18th century more flowery scents became fashionable and perfumers began to employ alcohol as a solvent, following the use in Cologne of a mixture of rosemary and citrus dissolved in alcohol. Alcohol-based extraction of natural oils from plants soon became a significant part of the local economy in towns such as Montpellier and Grasse in the south of France. As the perfume industry grew in Europe, scents were gradually differentiated according to gender, with floral fragrances considered to be feminine, while musk and other sharper scents were seen as masculine.

Perfumes were also seen as a way of repelling the miasma or foul air that was widely believed to cause illness (as Edwin Chadwick, who campaigned for improved sanitary conditions in 19th-century Britain, put it: 'all smell is disease'). Faced with outbreaks of sickness, people used perfumes to try and prevent infection. The 18th-century English writer Daniel Defoe described a church service during a plague outbreak: 'the whole church was like a smelling-bottle; in one corner it was all perfumes; in another, aromatics, balsamics, and a variety of drugs and herbs; in another, salts and spirits.' The modern interest in aromatherapy (the term was coined in 1925), which is based on the claimed effects of the scent of essential oils on well-being, can be traced back to early beliefs about the power of fragrance.

In the second half of the 19th century, perfume production became big business in Europe. Companies such as Rimmel and

Coty were set up; by the early 20th century, perfumes became intertwined with haute couture fashion, in particular through Coco Chanel and her striking No. 5 perfume. The link between perfume and desire—for sex, love, prestige, wealth, glamour, and uniqueness—is emphasized by advertising, and by the bottles in which the perfume is sold. Our wish to be associated with a brand, with a style, with a famous name, can seem more significant than the nature of the fragrance itself. The uses of perfume in a culture can change over a relatively brief time. In the 1980s in the United Arab Emirates a woman would not use perfume in public or when meeting men who were not family members. Nearly forty years later, a visit to a Dubai shopping mall will reveal both men and women using expensive and complex forms of scent and fragrance in public.

Perfumes can provide relief in the face of the most appalling suffering. Liana Millu, an Italian Jewish resistance fighter who was sent to Auschwitz—a place renowned for the smell produced by the awful conditions that the prisoners lived in and the stench of burning bodies from the camp crematoria—recalled how the arrival of a scented letter briefly transformed the prisoners' hellish lives:

> Of course everyone had to smell the perfume, so the note was passed around and sniffed ecstatically. I smelled it too—just the faintest breath of scent, as if the paper had been near face powder. Still, I pressed it close to my nostrils and inhaled greedily.

The politics of smell

Throughout history people have consistently used smell descriptors to present particular social groups, in particular immigrants or racial minorities, as 'other'. In the USA in the 19th century, Chinese migrants were accused of engaging in 'the foul-smelling sub-culture of opium smoking', while black slaves were alleged to 'stink damnably'; even long after emancipation, racist southerners claimed to be able to detect black people by smell. Politicians have used this kind of imagery to cultivate support for discrimination—in *Mein Kampf*, Hitler complained of the smell of the Jews, while in 1992 the French Prime Minister Jacques Chirac notoriously complained about the 'noise and smell' supposedly produced by Arab immigrants.

Travellers' accounts often reveal an appalled fascination with the smells of other cultures. At the beginning of the 16th century, the Dutch scholar Erasmus visited England and complained of the 'exhalation' that emerged from the floors, a bouquet that he suspected had incubated for decades and which included 'spittle, vomit, the urine of dogs and men, the dregs of beer, the remains of fish, and other nameless filth'. Around 400 years later, in *Aromatics and the Soul*, the Scottish surgeon Dan McKenzie quoted a fellow-countryman as saying, 'The East is just a smell! It begins at Port Said', before adding his own take, inadvertently revealing the powerful impact left on him by the olfactory delights of the exotic East: 'Who can ever forget the bazaar smells of India, the mingled must and fust with its background of garlic and strange vices, or the still more mysterious atmospheres of China with their deep suggestion of musk?' McKenzie's view of some of the smells he encountered in the British Isles had no such sensuous subtext, and he bitterly complained of 'the fumous and steamy stench of parboiled cabbage that filled the restaurant-car of the train for Belfast'. Perception of the smell of others is generally reciprocally negative—both sides of the social divide perceive the other as smelling unpleasant. During the British Raj in India, the Britons would complain of the smell of the Indians, while the Indians felt the same way about their imperial rulers.

The rich and powerful have always had near-exclusive access to expensive perfumes, scented baths, and, until very recently, toilet facilities. All of this led the rich to smell rather differently from the poor, with their inferior living and working conditions, and poor overall health. In 1709, a French perfumer suggested that different classes should use different fragrances, further cementing social divisions in the language of smell. Two centuries later, as George Orwell recalled bluntly in *The Road to Wigan Pier*, the children of the British ruling class were taught that 'the lower classes smell'. Given the odours that were prevalent in their houses and workplaces, this would hardly be surprising. In *The Condition of the Working-Class in England*, published in 1845, Friedrich Engels described how the typical English factory was poorly ventilated and was full of 'the smell of the machine oil, which almost everywhere smears the floor, sinks into it, and becomes rancid', while a decade earlier, in his novel *Père Goriot*, Balzac wrote of a Parisian boarding house as follows: 'It smells stuffy, mouldy, rancid; it is chilly, clammy to breathe, permeates one's clothing; it leaves the stale taste of a room where people have been eating; it stinks of backstairs, scullery, workhouse.'

This association of powerlessness and stench was sometimes complex—imperial purple, the rich hue created by a dye used by Roman aristocrats to colour their clothes, was made from the extract of the glands of marine rock snails and smelt terribly. However the fabric worn by the Emperor might have smelled, the hands of the dyers who created this material were said to stink of rotting fish—the Talmud generously granted the right of divorce to any married woman whose husband became a dyer. The most significantly smelly occupations in the growing cities of Europe were those associated with treating human and animal excreta—the sewermen who spirited the stuff away from cesspits underneath buildings and the tanners who would collect night soil for use in processing hides. In *Les Misérables*, Victor Hugo described 'these heaps of garbage at the corners of the stone blocks, these tumbrils of mire jolting through the streets at night, these horrid scavengers' carts, these fetid streams of subterranean slime'. Eventually, the great cities of Europe were obliged to find a new way of dealing with the tons of turds that were produced within their walls each day. In London the practice of chucking the stuff in the Thames led to the Great Stink of 1858, which was so overpowering that the House of Commons, situated on the banks of the river, had to be suspended. Within months, work had begun on an immense system of underground sewers, carrying the filth far away from the noses of Londoners, into the lower reaches of the Thames estuary.

One consistent form of discrimination that has been linked with odour is the oppression of women. In European cultures, witches were thought to smell, partly because they consorted with evil spirits and partly because they were women. There were three reasons why it was believed that prostitutes smelled—they were women, their sexual activity was associated with odours, and they often used perfume. Strikingly, one of the words for 'prostitute' in Romance languages—*pute* in French, *puta* in Spanish—has the same root as 'putrid'. From the ancient Greeks to Christian theologians, many people (most of them men) have claimed to be able to detect whether a woman is a virgin on the basis of her smell. In many ancient religious ceremonies, the bride was heavily perfumed, both as a way of hiding her supposedly naturally foul odour and as a way of repelling evil spirits, djinns, or demons. This is just one example of the use in many cultures of fragrances to mark rites of passage—puberty, marriage, death, and so on.

The use of perfumes by women is often seen as a form of seduction, and therefore both attractive but also problematic, perhaps because of its association with prostitution and because it suggests that it provides women with an additional source of power and influence over men. In 1770, a law was passed in England that stated that any woman who used 'scents, paints, cosmetic washes' to 'seduce and betray into matrimony any of His Majesty's subjects' would be subject to the law against witchcraft.

The ethnography of smell

Different cultures perceive the world of smell in different ways and provide a fascinating glimpse into other ways of smelling.

A 17th-century dictionary of Quechua, the language of the Inca, reveals eight different terms for smelling—smelling something bad, smelling together, smelling another person, and so on—indicating the importance of olfaction in this culture. Contemporary anthropologists observing other societies can encounter alternative smell-worlds that to western noses are quite unusual. For example, the Desana people of Columbia, who call themselves 'wira' ('the people who smell'), use olfaction to move about the forest, identifying animals such as agouti and jaguar by their smells; they even consider that their own odours mark out their territory. According to the Desana, men and women smell differently, with women smelling like ants or worms. So significant is smell in the Desana world-view and culture that they believe that smells are detected with the whole body, not the nose alone.

The Ongee people, who live on the Andaman Islands in the Bay of Bengal, are highly attuned to smells—when they greet each other they say 'How is your nose?' They measure the passage of weeks and months through what is effectively a smell calendar, based on the plants that bloom each season, as each plant produces a different 'aroma-force'. This floral timetable has a practical aspect—wild honey produced at different times of the year has varied flavours, as the bees visit differently scented flowers. The Ongee also identify parts of the islands according to their scents and use smell in their belief system, according to which smells can be used to attract the odourless spirits of the dead. Another forest-dwelling people, the Umeda of New Guinea, are extremely sensitive to odours such as the smell of smoke, or the scent of a possum—things that cannot be seen in the dense jungle can be smelled. Elsewhere, the Dassanetch of Ethiopia use the smells of livestock and farming practices such as field-burning to identify the seasons. Given the significance of cattle in their culture, the Dassanetch cover themselves with the smell of cow milk, fat, urine, and faeces, and women use the smell of cow to make themselves attractive.

Many of these peoples have ways of classifying odours that are unlike those prevalent in the west, each based on local ecologies and culture. However, the social dynamics underlying these classifications generally show remarkable similarities, with unpleasant odours used to describe those lower on the local social scale. For example, the Serer Ndut people of Senegal consider that Europeans and monkeys smell 'urinous', whereas the Serer Ndut themselves are at the other end, smelling 'fragrant'; neighbouring tribes are somewhere in the middle, smelling 'fishy' (Table 3). Strikingly, among the most attractive odours, along with that of the Serer Ndut themselves, is the smell of raw onions, which many in the west would not appreciate. This emphasizes the culture-bound nature of most of our interpretations of smell.

Researchers recently studied how twenty different languages from around the world represent perceptual experiences. They found that there was no universal hierarchy of the senses—for example, although English reserves its most precise terms for colour, in other languages such as Lao (Laos), Farsi (Iran), and Tzeltal (Mexico), taste had the most specific terms. Hunter-gatherers such as the Umpila from Australia used more precise terms for smells than they did for colours. This backed up previous research that found that Jahai hunter-gatherers in Malaysia and Thailand identified odours as easily as they did colours; indeed, in a comparison of Jahai and Dutch people, the Jahai identified an odour verbally within three seconds, while the Dutch took over five times as long.

To explore this apparent link between human ecology, perception, and language, Professor Asifa Majid, now of York University, studied the olfactory abilities of two groups of people living in the humid and gloomy tropical rainforest of the Malay Peninsula. These groups—the hunter-gatherer Semaq Beri and the horticulturalist Semelai—speak closely related languages but differ in their mode of subsistence. When tested with a range of odours, the hunter-gatherers found it easier to name odours than

Table 3 Olfactory classification system of the Serer Ndut people of Senegal

1. Urinous	Europeans
	monkeys, horses, dogs, cats
	plants used as diuretics, squash leaves
2. Rotten	cadavers
	pigs, ducks, camels
	creeping plants
3. Milky or fishy	nursing women, neighbouring tribes
	goats, cows, antelopes, jackals, fish, frogs
4. Acidic	spiritual beings
	donkeys
	tomatoes, certain trees and roots
5. Fragrant	Serer Ndut, Bambara
	flowers, limes, peanuts, raw onions

From: Classen, C., Howes, D., and Synnott, A. (1994) *Aroma: A Cultural History of Smell* (London: Routledge).

did the farmers. Furthermore, in Semaq Beri culture, personal odour is considered to be extremely important, with social spaces being managed to avoid odour mingling. Despite our western prejudices, and the power of English terms like 'to see' meaning 'to understand', human senses are not necessarily organized with vision at the top and smell at the bottom. Much of how we think and speak about smells is culturally contingent.

The smell of culture

Although scent is evidently significant in so many aspects of so many cultures, until recently historians and intellectuals paid little attention to a sense that, in western culture at least, was considered to be of lesser significance than vision and hearing.

All that changed in a little more than a decade, with the publication of two landmark books. In 1982 the French historian Alain Corbin published an account of smell in 18th- and 19th-century France, translated in 1996 as *The Foul and the Fragrant*. Beginning in the 1780s, with the country on the brink of revolution, Corbin traced the links between public health and olfactory perception. Then in 1994, Constance Classen, David Howes, and Anthony Synnott published *Aroma: The Cultural History of Smell*, which explored smell across time and space using anthropological and cultural data as well as historical insights (some of the examples in this chapter are taken from this book). Since then, many researchers have explored the role of odours in specific parts of history and culture, such as the history of Christianity, the history of the perfume industry, or the link between smell and flavour in world cuisines.

References to smell can provide striking imagery that powerfully evokes a moment in time to the reader. In my book on the liberation of Paris in August 1944, I reproduced the account of the photographer Lee Miller, who arrived in the French capital just after the fighting was over and was immediately struck by how the smell of the city had changed:

> It used to be a combination of patchouli, urinals and the burnt castor oil which wreathed the passing motorcycles. Now it is air and perfume wafting across a square or street. All the soldiers noticed the scent and, asked what they thought of Paris, became starry-eyed. They said, 'It's the most beautiful place in the world and the people smell so wonderful.'

Literature can provide a particularly powerful impression of how people view—or rather smell—the olfactory world. As the British academic John Sutherland has pointed out, the work of George Orwell often uses olfactory imagery, from the opening passages of *1984* ('The hallway smelt of boiled cabbage and old rag mats'), to the smell of a house concubine in *Burmese Days*

('A mingled scent of sandalwood, garlic, coconut oil and the jasmine in her hair floated from her'). For Orwell, odour was an essential component in describing a scene and conveying atmosphere. Not all writers are so sensitive to scent—there are only a handful of references to odour in Hemingway's writings, and only one in all of Jane Austen's work (in *Mansfield Park*).

In 19th-century Europe, novelists such as Zola, Balzac, and Dostoevsky all used olfactory imagery to convey place in their writings, while for the French novelist Huysmans and the poet Rimbaud, smells were linked to sensuous, sexual feelings. In *Ulysses*, written by James Joyce in 1918–20, we are drawn into the mind, mouth, and nose of the central character Lionel Bloom when we learn that 'he liked grilled mutton kidneys which gave to his palate a fine tang of faintly scented urine'. In Aldous Huxley's 1932 dystopian novel *Brave New World*, smell plays a primordial role—'scent taps' distribute fragrances, specific odours are used to condition the different genetically determined classes, while a 'scent organ' is used to present works of olfactory art that mix sandalwood, new-mown hay, 'a whiff of kidney pudding', and 'the faintest suspicion of pig dung'. Odour is used by Huxley to contrast the contrived and artificial world of civilization and the grubby, but infinitely more real world of the 'savages'.

Probably the best-known European novel to deal with smell is Patrick Süskind's *Perfume*, which appeared in 1985 and has sold over 20 million copies (key aspects of the olfactory plot bear a striking resemblance to Roald Dahl's 1974 short story *Bitch*). Set in 18th-century France, *Perfume* contains some extremely evocative passages describing the smells of Paris and of the foothills of the Alps around Grasse. The central character, Grenouille, has an extraordinarily sensitive nose; coupled with his obsession with capturing the smell of one particular young woman, this leads to tragedy. In other cultures, where fragrance has more precise significance, odour has played a consistent role in fiction. In what is arguably the world's first novel, *The Tale of*

Genji, written in 11th-century Japan by a noblewoman, Murasaki Shikibu, the fragrances used in the home (*soradakimono*) play a key role, acting as metaphors for the characters as well as highlighting emotions and the passing of the seasons.

There have been repeated attempts to use smell in exhibitions and art, although nothing on the scale of Huxley's scent organ has yet been built. The Jorvik Centre in York, which opened in 1984, contains a display in which visitors are transported through a diorama of the area as it was a little over 1,000 years ago, peopled by animatronic Viking villagers. You are drawn into the experience by the use of smells, mainly foul but some fragrant, which correspond to the various scenes. The centre is one of the most popular attractions in the UK, not least because of its embracing of the olfactory.

Imaginative directors have tried to use smell in film, such as in the 1929 musical *The Broadway Melody*, during which one cinema allowed perfume to drift down from the ceiling. This innovation was not entirely original—two millennia earlier, theatrical productions in ancient Rome would spray fine jets of saffron-scented wine above the audience at appropriate moments in the plot. Walt Disney considered using smells to accompany his 1940 cartoon masterpiece, *Fantasia*, before abandoning the idea due to cost. In the 1950s two competing systems were developed for using smell in cinema—Smell-O-Rama and Smell-O-Vision. Neither was a financial or an artistic success. They both suffered from the same problems—delivering the odours to the viewer, removing them rapidly so they did not linger, and choosing fragrances that would be perceived in the same way by everyone in the audience. Perhaps a complex system is not needed and the power of suggestion and the strength of smell in our mental life could suffice—on 1 April 1965 the BBC televised an interview with a man who claimed to have invented a technique called Smellovision that would broadcast odours to the viewer (or smeller). As the screen showed footage of coffee brewing and onions being chopped, viewers all over the country, who had not noticed the date, reported they could smell the odours. Whatever the truth of this story, olfactory suggestion is certainly at work in the 2020 Oscar-winning film *Parasite*. In this Korean satire, smell plays a key and powerful role in the plot, without any trace of Smell-O-Rama.

In the 1970s, micro-encapsulation allowed smells to be released at will by scratching a patch of paper impregnated with tiny bubbles of fragrance. This 'scratch and sniff' technology led to a brief fashion for linking smells to content in a variety of cultural products, such as video games, books (including a children's series called Smelly Old History, with titles such as *Roman Aromas* and *Victorian Vapours*), comics, television series, and even pornography. John Waters's 1981 film *Polyester*, which was filmed in 'Odorama', involved a card with ten patches that had to be scratched at appropriate points. In keeping with the deliberately outrageous nature of the film, the smells included skunk, dirty shoes, and faeces. More wholesome applications of fragrance in film can be smelled in various rides at Disneyland resorts, where aromas such as pie, orange blossom, or watermelon are wafted over visitors at appropriate points.

Advertisers have primarily used scratch and sniff in the most obvious way—to market perfume. However, in February 1989 the US weapons manufacturer BEI Defense Systems published a full-page advert in *Armed Forces Journal International*, showing a BEI-manufactured Flechette rocket destroying an enemy helicopter. The page was impregnated with microcapsules containing the odour of gunpowder, and the slogan used was 'The smell of victory'. The ad knowingly riffed on a phrase from Francis Ford Coppola's 1979 hallucinatory film about the Vietnam war, *Apocalypse Now*. In a famous scene, the psychopathic Lieutenant Colonel Kilgore (played by Robert Duvall) proclaims: 'I love the smell of napalm in the morning. You know, one time we had a hill bombed for 12 hours. When it was all over, I walked up. We didn't find one of 'em…The smell, you know that gasoline smell, the whole hill. Smelled like victory.'

The smell of cities

It is generally thought that throughout the 20th century, western culture became increasingly deodorized. In the 18th century, philosophers such as Kant and thinkers such as Buffon had argued that the sense of smell had something of the animal about it, and that it was not part of the finer, aesthetic sense that humans possess. This intellectual disdain for smell was soon accompanied by growing deodorization. With the cleaning up of cities through increased sanitation, followed in the 20th century by clean air and a decline in industry, our cities smelled less (in the UK, bad smells became legally enshrined as a 'nuisance' in a law of 1875).

At the same time there was a growing focus on personal hygiene, building on paranoid fears about having bad breath or body odour (it was called BO in adverts—so bad it could not even be named), first aimed at women and then at men (Figure 23). There have even been repeated and misguided attempts to persuade women to use vaginal deodorants and scented sanitary products. Bad personal odour was seen as reprehensible and leading to a lack of professional or amorous success. This was reinforced by advertising—Listerine was initially an antiseptic, but after being marketed as a mouthwash focusing on removing bad breath, sales rocketed and the company's profits soared forty-fold in the space of seven years. Even where odours have become important in western society, this often relates to cleanliness. In the 1960s, manufacturers discovered that merely adding lemon fragrance to a detergent led customers to perceive it as more effective at cleaning, although the formula had not changed. The same effect has been noted for shampoos, which are perceived as producing more suds, simply by adding a fragrance.

23. Advertisement for Lifebuoy Soap from the 1930s.

Despite the cultural trend towards deodorized uniformity, sectors of every city still have their unique odour—parts of Dublin smell of burning peat, the Paris metro still has its own ineffable whiff, the seaplane dock in Vancouver smells of a gorgeous blend of ozone and kerosene, some streets in Sofia in the late spring have the strong spermy smell of clematis, while inner-city breweries such as the one not far from my house regularly send the sharp smell of hops into the air. These kinds of local variations have been used by the sensory designer and smell collector Kate Maclean to create beautiful smell maps of different locations around the world. Architects and urban planners have recently taken an interest in this question, with a significant role played by my good friend and colleague the late Victoria Henshaw, to whom this book is dedicated. Victoria went on public 'smellwalks', which she organized in cities all over the world. She would visit urban landscapes and encourage participants to pay attention to the smells that surrounded them, from throaty diesel pollution to the nauseating tang of a urine-spattered back alley or the thick reek of old fat congealed on restaurant extractor fans. As she put it, her aim was to focus on 'the fetid and foul, the tantalising and satisfying, the familiar and surprising'. Researchers have even applied her approach to reconstructing the smell landscape of the area around the Roman fort at Vindolanda, on Hadrian's Wall. One of the key lessons for the future that is embodied in Victoria's work is that there are ways of planning for smell in cities, allowing inhabitants to encounter smells, including installing fragrant plants to act as an olfactory buffer to roads, or creating sensory gardens that can delight the public and can even act as a site for helping to stimulate people with dementia.

The significance of smell in the built environment extends to the most artificial and unearthly locations humans have yet encountered. The Apollo astronauts reported that the Moon smelled of spent gunpowder or like a fire that had just been put out—the fine moondust on their suits came off inside the lunar lander, spreading the smell of powdered volcanic rock. Similarly, after returning to their spacecraft from a spacewalk, astronauts have reported the smell of frying, while the International Space Station is said to smell of hot oil—both effects seem to be due to hydrocarbon molecules in space being transported to the inside of the vehicle. Future space smells may not be quite so pleasant. The surface of Mars consists largely of iron, magnesium, and sulphur, suggesting that when we eventually visit the red planet, the dust on our suits will make our spaceships and habitations smell of farts.

Chapter 7
The smell of the future

The single most salient scientific fact of the current age is our recognition of a climate emergency, caused by humanity's massive release of CO_2 into the atmosphere. Our future will be determined by this crisis and how we respond to it. Although the most significant aspect of this emergency is an increase in temperature and consequent climate change, shifts in temperature and increases in the CO_2 concentration of the surrounding medium (air or water) will have an impact on what there is to smell and how organisms smell it.

Marine organisms will be particularly susceptible as dissolved atmospheric CO_2 increases the acidity of the oceans, affecting both the odours in the water and the way that receptors function. Gene function in fish olfactory neurons is affected by increased CO_2 levels, with some genes becoming more active, others less so—these changes lead to decreased neuronal growth and plasticity, and to changes in neurotransmitter activity. As a result, the way that odours are detected in the fish's nose is altered by the levels of CO_2 in the water. The same is true at the next stage up, in the brain—again, changes to gene function lead to reductions in neuronal excitability and plasticity. This means that as the CO_2 levels in the atmosphere increase and aquatic habitats become more acidic, fish brains will be less able to respond to odours and may find it harder to learn the significance of smells. This effect of increased acidity on olfaction will not only be a problem in the oceans—freshwater studies by researchers in Germany have shown that increased dissolved CO_2 reduces the ability of *Daphnia* populations to grow anatomical defences when they smell a predator.

For the moment, little is known about how increased CO_2 levels will affect the olfactory function of terrestrial organisms. They may similarly affect receptor function and directly disrupt olfactory ecology, given the role the gas plays in the life cycle of many terrestrial animals, which use CO_2 to track prey, to identify breeding sites, or as an alarm cue. The consequence of higher CO_2 levels on temperature will also have a broader effect on both general odour detection and pheromonal systems. Higher temperatures lead to increased odour volatility, allowing smells to be detected over longer distances, and at different times of day and in different seasons. The consequences for animals and plants are unknown, but the overall effects will probably include sudden, disruptive changes to olfactory communication.

Increased CO_2 is not the only element in the current crisis. Plastic pollution is having a drastic impact on marine birds, through its olfactory effects. Many pelagic birds, such as shearwaters or petrels which forage in the deep ocean, are highly attracted to dimethylsulphide (DMS), a gas which is released when krill—tiny planktonic crustaceans—graze on phytoplankton. This volatile gas floats into the atmosphere and can be detected by the birds, which use it as an indicator of the location of their preferred crustacean food—experiments have shown that you can attract these birds by pouring small quantities of DMS onto the ocean. When plastic enters the marine environment, the action of sea water releases DMS from the surface of the plastic, attracting birds, which then eat the debris they find in the ocean. One study reported that around half of the birds from DMS-responsive species had ingested plastic, compared to less than one in ten birds from species that could not detect the gas. This tragic confusion partly explains the distressing images of dead birds with stomachs full of bits of plastic garbage, the detritus of our civilization.

Similar effects are seen for fish such as anchovies, which are highly attracted to plastic that has been in the ocean for some time and has a layer of bacteria growing on it. The exact chemicals that attract the fish are unclear, but they may be part of the complex cycle involving DMS. Whatever the case, the plastic we use so heedlessly, which generally finds its way into the oceans, is having a devastating consequence on the survival of many sea-faring organisms by interfering with their natural olfactory responses and ecology.

Things are not much better on land. Guy Poppy's research group at Southampton University has shown that diesel pollution, and in particular mono-nitrogen oxide, is altering the behaviour of pollinators such as bees, by directly transforming the fragrances of some flowers. Some components of these bouquets can be reduced by diesel exhaust, others can be entirely erased, and, most alarmingly, some odour molecules can have their orientation altered, like turning a left glove into a right glove. In the laboratory, these changes affect the ability of bees to learn to associate a fragrance with a reward, suggesting that pollution-induced changes to the olfactory link between flowers and insects may be affecting the survival of both sets of organisms.

The effect of air pollution on human olfaction is well known. People living in areas with high levels of pollution from industry or traffic have decreased olfactory function; this is particularly the case in children. Olfactory dysfunction increases from around 2 per cent in rural, non-polluted areas to 10 per cent in urban areas with high levels of air pollution. A major factor in the observed differences in olfactory sensitivity between Europeans from urban areas and various indigenous peoples living in relatively unspoilt environmental conditions is thought to be air pollution. Reduced olfactory function can have a series of complex effects on emotions and cognition, as well as mental health, including through the loss of much of the sense of taste, which can be a factor in depression. There have been suggestions that air pollution may be linked with neurodegenerative forms of dementia—exposure to air pollutants certainly increases the likelihood of damage to our olfactory neurons.

Unsurprisingly, pollution does not affect all social groups equally—in this area as in others, there is a close link between health and class. This in turn often overlaps with ethnic and educational differences, which are underpinned by class. Similar effects can be observed for the other senses, in particular noise pollution. This has led Kara Hoover of the University of Alaska to highlight the existence of sensory inequities—variations in the sensory environment that are linked with class. In the case of smell, unless drastic action is taken to curb and reduce air pollution, in particular in inner cities and around industrial sites, these differences will increase over the coming decades with unknown implications for public health and well-being.

People who are already oppressed may find their situation worsened by their olfactory surroundings. Deborah Davis Jackson of Earlham College in the USA has worked with members of the Aamjiwnaang band, a Canadian First Nation who live on a reserve in Ontario. The place was once full of the smells of honey scented clover flowers, fragrant sassafras, and wild ginger, and sap would be harvested from maple trees when the people could 'smell spring'. Since the middle of the 20th century, a large number of chemical and petroleum plants have been built nearby and the region is now known as 'chemical valley'. The elders call the place Winaaptae, which means 'it is blowing dirty', and claim that 'each corner of the reserve has its own special stench', inducing anxiety and fear of the effects of pollution. As the population has slowly dwindled, the foul odours that permeate the place have added to the brutal deprivation and sense of alienation that these people suffer.

For the moment, we know very little about the complex consequences of the changes we are wreaking on the planet, and on ourselves. These examples of how smell will be affected by changes to the ecosystem in the coming decades undoubtedly only scratch the surface. They highlight the need both for more research to fully understand what is happening, and above all for action to turn back the rising tide of CO_2, plastic, and pollution while there is still time.

Artificial noses

Visions of the future generally involve advanced technology that may, in the words of Arthur C. Clarke, be indistinguishable from magic. While implants for hearing-impaired people are now routine, and some progress has been made on artificial vision, there is no immediate prospect of an artificial nose interfacing with our brain. The first electronic nose was reported in 1982 by my friend and colleague Krishna Persaud, together with George Dodd, both then at the University of Warwick. Using three different metal oxide sensors that changed their conductivity on contact with different gases, they were able to crudely mimic the ability of the mammalian nose to distinguish a number of odours.

Over the last four decades there has been enormous interest in building artificial noses, with varying degrees of success. Instruments designed to detect specific odours have been deployed in a variety of areas, including sewage treatment, food security, and medicine. These techniques generally use a relatively limited number of sensors that detect only a few odours of interest—paradoxically, this highly specialized sensitivity is what underlies their usefulness. Like the hedgehog in the ancient Greek proverb, they do one thing very well. For example, there are a range of commercial devices that help physicians diagnose urinary tract infections, distinguishing the pathogen responsible for the infection by detecting the characteristic volatile organic compounds it produces.

Following from the observation that dogs can apparently detect patients with bladder cancer by sniffing urine samples, there has been growing interest in using electronic noses to detect other forms of the disease, including prostate cancer and colorectal cancer. Other diseases may also be diagnosed by smell in the future. Working with a woman with a very sensitive nose who claimed to know if an individual suffered from Parkinson's Disease based purely on their odour, a group of researchers led by my colleague Professor Perdita Barran has recently been able to identify a number of molecules in the sebum of affected patients. These compounds may be biomarkers for this disease and could lead to earlier diagnosis.

The difficulty in building more broadly sensitive electronic noses is a result of two problems: finding an appropriate sensor material that differentially responds to odours and transduces that change into a detectable signal, and the difficulty of then forming those signals into a precise tool for identifying a wide range of odours (in other words, doing what even the simplest animal's olfactory system can do quite effortlessly). Many sensors have been tried, including exotic materials such as nanowires, quantum dots, and graphene, and increasingly researchers have been attempting to fuse these materials with organic molecules such as odorant binding proteins, MUPs, or olfactory receptors.

While such bionic noses seem theoretically straightforward—the organic components are naturally 'tuned' to detect particular odour molecules—connecting the organic and inorganic molecules, and using the conformation changes in the organic elements to produce a detectable signal, is proving highly challenging, although there have been some small successes in the laboratory. A linked problem involves developing algorithms that can reliably use the output of such sensors to identify odours with the same degree of accuracy as your nose (Figure 24). Both deep learning programmes and biomimicry, using direct applications of our growing understanding of how even simple brains construct

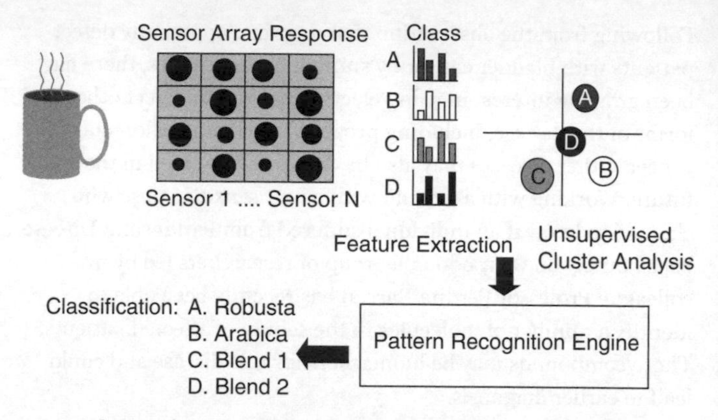

24. A schema of an electronic nose for detecting blends of coffee. Sensor arrays produce responses to key components of the blend, which are represented as histograms. A computer program then analyses these to produce clusters, which are compared with previously identified patterns by a pattern recognition engine which then identifies the blend.

an image of the olfactory world, will probably be involved in this decisive but complex step.

Future organisms

The significance of smell in the ecology of a large number of pest organisms and disease vectors has led a number of researchers to consider altering these systems in order to disrupt harmful activity. One project, linked to agricultural pests, has focused on the aphid alarm pheromone. When aphids are attacked by predators they secrete this substance, leading neighbouring aphids to rapidly disperse. This is a very short-lived pheromone which is detected by a single class of olfactory receptor in sensilla on one particular segment of the aphid's antenna. Researchers at Rothamsted Research, led by John Pickett, hatched the cunning plan of introducing into a strain of wheat a synthetic gene that would produce the aphid alarm pheromone. The crop would literally smell scary to the aphids and keep them away. In the laboratory, modified wheat was indeed able to produce the pheromone and aphids spent less time in airstream from the manipulated plants, while parasitoids of aphids showed increased levels of foraging on the plants, presumably eavesdropping on the pheromone. But in a field trial carried out in 2012–13 on a 36 m^2 plot, there was no significant change in aphid numbers or in the activity of the parasitoids on the modified crop. Despite the obvious potential shown by this system, it did not work in the real world. One of the problems may have been that while in nature the pheromone is released in pulses when the aphids are attacked, the manipulated wheat released it continuously. The aphids may simply have got used to it; more sophisticated ways of regulating pheromone production may be necessary.

This kind of approach will become easier to implement with the arrival of CRISPR gene editing technology. Virtually any gene, in any animal or plant, can now be changed in a way desired by scientists, and without involving the introduction of any foreign DNA (an issue that preoccupies many anti-GMO campaigners). One area of particularly active research relates to altering mosquitoes and other disease vectors. For example, if the domestic form of an *Aedes aegypti* olfactory receptor was indeed responsible for the shift in the fly's host preference from forest mammals to humans, it would eventually be possible to create something called a gene drive, which would rapidly spread throughout the population of mosquitoes, changing the domestic form of the gene to the forest form, thereby leading to a reduction in the number of humans who are bitten. Substantial ecological and ethical investigations and strict international regulation and agreement would be necessary before this kind of experiment was carried out—mosquitoes can travel long distances, in particular in pools of water on ships, and a well-meaning decision taken in one part of the world could rapidly have unexpected consequences in another.

Recovering smell

As many readers will know, either from their own experience or from that of family members or close friends, losing one's sense of smell through disease, injury, or simply old age can be a catastrophic experience. Without a sense of smell people may feel isolated and depressed, including losing sexual desire, and their interest in food can decline as their sense of taste becomes reduced to the key taste modalities. People with phantosmia—smelling things that are not there—can find the condition deeply distressing, particularly if the phantom odours are repellent. Despite the seriousness of these effects, if you go to your family doctor with anosmia or phantosmia you may be told there is nothing that can be done and you will have to learn to live with it. This is because although olfactory neurons form a stem cell network that can regrow, in many cases either the regrowth does not happen or the new neurons are not able to find their way to the correct glomerulus in the brain. This therapeutic deficit may be filled in the coming years, in a number of ways.

First, physicians are becoming increasingly aware of the possibility of alleviating some aspects of acquired anosmia through what is called smell training. Developed by Thomas Hummel of the Interdisciplinary Centre for Smell and Taste in Dresden, smell training involves daily smelling of a number of different fragrances, such as rose, lemon, clove, and eucalyptus, for up to a year. This can help speed up the slow spontaneous recovery that occurs in many cases—around 20 per cent of people who have lost their sense of smell after head trauma and over 50 per cent of people with anosmia following an upper respiratory tract infection eventually recover olfactory function. Ear, nose, and throat surgeons are beginning to adopt this approach, and if you or someone close to you has lost their sense of smell, you should ask your doctor about this.

Much further distant is the possibility of following the successful use of auditory implants to alleviate deafness. This could be done through cell transplants of olfactory epithelium, or even, ultimately, the development of bionic noses. However, the complexity of the olfactory system compared to hearing means that this is at least decades away, and it may never be realized. Another option would help the subset of patients whose smell disorders are caused by defective cilia on their olfactory sensory neurons—it may be possible to improve cilia function through gene therapy. Again, these are distant prospects for the moment.

As part of the growing awareness of both the significance of smell loss and the possibility of alleviating some of the symptoms, a number of patient groups have been set up. These include FifthSense.org.uk, a UK-based charity that aims to provide support to people with smell and taste disorders across the globe, and abscent.org, which works closely with physicians and patients to raise public awareness of smell loss and help make a difference to those with the condition. There is no magic solution to smell loss or to phantosmia, but there is no need to suffer in silence—sharing experiences and exploring potential ways of alleviating the condition can be life-changing.

In conclusion

The sense of smell is rich and complex, affecting everything from our deepest emotions and inner mental life to the astonishing interactions that underpin the global ecosystem. Over the last three decades our understanding of how smell works has been transformed, in particular following the Nobel Prize-winning work of Linda Buck and Richard Axel, who identified the mammalian genes that produce olfactory receptors. But despite the vast amount of work by scientists around the world, studying a whole range of animals, our understanding of the mechanisms of smell, in particular how we form olfactory images in our mind, remains rudimentary. The main challenges for the future are to discover exactly what happens when an odour binds with a receptor, how receptor neurons respond differently to a wide range of odours, and what is involved in the pattern of activity that is created in our brains when we smell. In particular, researchers need to focus on what happens when we smell real smells, which are generally complex blends that are detected within a rich olfactory environment. There are a number of different models to explain what happens when we and other animals smell scent mixtures—to put it simply, blends are either sensed as the sum of their parts, or as something novel. It is quite possible that different blends are sensed in different ways by the same species, or that the same blend is processed in different ways by different animals.

Part of the problem with making progress in understanding smell is that so much of our scientific comprehension of sensory processing is based on the visual system. Olfaction is very different, involving highly complex stimuli and very different neuronal structures; the predominance of concepts and even words associated with vision represent a limit on our attempts to understand what is happening. Some researchers have recently begun to explore the philosophical implications of these problems—Andreas Keller's 2016 book *Philosophy of Olfactory Perception*, and Ann-Sophie Barwich's 2020 *Smellosophy* both explore the similarities and differences between smell and the other senses, in particular what it is that we smell when we smell. Building on the work of the philosopher Clare Batty, they both conclude that we do not simply perceive chemical objects; instead we carry out a perceptual categorization based on chemical stimuli and heavily modulated by experience.

This might seem a subtle nuance, but it helps explain the magical aspect of smell, how odours are so often attached to meaning, and how the same odour can have very different meanings to different people. In this sense, the olfactory code represented by the pattern of activity in receptor neurons and glomeruli is merely one aspect of our perception. That initial pattern is nuanced and modulated by top–down, contextual information that includes salience and value, and which reverberates in memory. This kind of thing is not limited to humans, it is also taking place, at a more rudimentary level, in every animal on the planet.

In the coming decades smell, for so long overlooked by philosophers, scientists, historians, and indeed by many of us in our everyday lives, will transform not only our understanding of what it is to be human, but even more significantly, how we experience the natural world and how the whole ecosystem functions.

In 1985, the American physician Lewis Thomas published a brief essay entitled 'On Smell', in which he poetically explored this most powerful sense, and pondered on one of the smells that could work its magic on his mind and which he could conjure up at will—the smell of smoke. 'Tobacco burning, coal smoke, wood-fire smoke, leaf smoke. Most of all, leaf smoke.' The scent of leaf bonfires, he feared, with its 'aroma of comradeship', was endangered in the modern world; if it were to disappear, part of us would disappear, too.

All of us who are able should try to gain increased pleasure, and insight into the olfactory world that surrounds us, by deliberately exploring this special, magical sense, rather than taking it for granted. As Thomas's example shows, precious, evocative smells can be found in the most prosaic of circumstances, surrounding us without us even noticing it. But we can change that, by paying attention and deliberately seeking out scents. You can start right now—put your nose into the pages of this book and smell deeply. Go on!

References

Only some of the relevant references are given here, due to space constraints. I am happy to supply details of any research referred to in the text that is not listed below.

Chapter 1: How we smell

Andersson, M. M., Löfstedt, C., and Newcomb, R. D. (2015) Insect olfaction and the evolution of receptor tuning. *Frontiers in Ecology and Evolution* 3:53.

Bear, D. M., et al. (2016) The evolving neural and genetic architecture of vertebrate olfaction. *Current Biology* 26:R1039–49.

Berck, M. E., et al. (2016) The wiring diagram of a glomerular olfactory system. *eLife* 5:e14859.

Bushdid, C., et al. (2014) Humans can discriminate more than 1 trillion olfactory stimuli. *Science* 343:1370–2.

Heydel, J.-M., et al. (2013) Odorant-binding proteins and xenobiotic metabolizing enzymes: implications in olfactory perireceptor events. *The Anatomical Record* 296:1333–45.

McGann, J. P. (2017) Poor human olfaction is a nineteenth century myth. *Science* 356:eaam7263.

Malnic, B., et al. (1999) Combinatorial receptor codes for odors. *Cell* 96:713–23.

Padmanabhan, K., et al. (2019) Centrifugal inputs to the main olfactory bulb revealed through whole brain circuit-mapping. *Frontiers in Neuroanatomy* 12:115.

Weiss, T., et al. (2020) Human olfaction without apparent olfactory bulbs. *Neuron* 105:35–45.

Chapter 2: Smelling with genes

Araneda, R. C., et al. (2000) The molecular receptive range of an odorant receptor. *Nature Neuroscience* 3:1248–55.

Benton, R., et al. (2006) Atypical membrane topology and heteromeric function of Drosophila odorant receptors in vivo. *PLoS Biology* 4:e20.

Buck, L. and Axel, R. (1991) A novel multigene family may encode odorant receptors: a molecular basis for odor recognition. *Cell* 65:175–87.

Butterwick, J. A., et al. (2018) Cryo-EM structure of the insect olfactory receptor Orco. *Nature* 560:447–52.

Greer, P. L., et al. (2016) A family of non-GPCR chemosensors defines an alternative logic for mammalian olfaction. *Cell* 165:1734–48.

Hoover, K. C., et al. (2015) Global survey of variation in a human olfactory receptor gene reveals signatures of non-neutral evolution. *Chemical Senses* 40:481–8.

Hughes, G. M., et al. (2018) The birth and death of olfactory receptor gene families in mammalian niche adaptation. *Molecular Biology and Evolution* 35:1390–406.

Jones, W. D., et al. (2005) Functional conservation of an insect odorant receptor gene across 250 million years of evolution. *Current Biology* 15:R119–21.

Keller, A., et al. (2007) Genetic variation in a human odorant receptor alters odour perception. *Nature* 449:468–72.

Liberles, S. D. and Buck, L. (2006) A second class of chemosensory receptors in the olfactory epithelium. *Nature* 442:645–50.

Matsui, A., et al. (2010) Degeneration of olfactory receptor gene repertories in primates: no direct link to full trichromatic vision. *Molecular Biology and Evolution* 27:1192–200.

Trimmer, C., et al. (2019) Genetic variation across the human olfactory receptor repertoire alters odor perception. *Proceedings of the National Academy of Sciences USA* 116:9475–80.

Chapter 3: Smell signals

Greene, M. J. and Gordon, D. M. (2003) Social insects: cuticular hydrocarbons inform task decisions. *Nature* 423:32.

Jallon, J.-M. (1984) A few chemical words exchanged by Drosophila during courtship and mating. *Behavior Genetics* 14:441–78.

Karlson, P. and Lüscher, M. (1959) 'Pheromones': a new term for a class of biologically active substances. *Nature* 183:55–6.

Keller, L. and Nonacs, P. (1993) The role of queen pheromones in social insects: queen control or queen signal? *Animal Behaviour* 45:787–94.

Li, Y. and Dulac, C. (2018) Neural coding of sex-specific social information in the mouse brain. *Current Opinion in Neurobiology* 53:120–30.

Murata, K., et al. (2014) Identification of an olfactory signal molecule that activates the central regulator of reproduction in goats. *Current Biology* 24:681–6.

Savarit, F., et al. (1999) Genetic elimination of known pheromones reveals the fundamental chemical bases of mating and isolation in Drosophila. *Proceedings of the National Academy of Sciences USA* 96:9015–20.

Schaal, B., et al. (2003) Chemical and behavioural characterization of the rabbit mammary pheromone. *Nature* 424:68–72.

Swammerdam, J. (1758) *The Book of Nature* (London: Seyffert).

Van Oystaeyen, A., et al. (2014) Conserved class of queen pheromones stops social insect workers from reproducing. *Science* 343:287–90.

Chapter 4: Smell, location, and memory

Dahmani, L., et al. (2018) An intrinsic association between olfactory identification and spatial memory in humans. *Nature Communications* 9:4162.

Eichenbaum, H., et al. (1984) Selective olfactory deficits in case H.M. *Brain* 106:459–72.

Grillet, M., et al. (2016) The peripheral olfactory code in Drosophila larvae contains temporal information and is robust over multiple timescales. *Proceedings of the Royal Society B* 283:20160665.

Jacobs, L. F. (2012) From chemotaxis to the cognitive map: the function of olfaction. *Proceedings of the National Academy of Sciences USA* 109:10693–700.

Louis, M., et al. (2008) Bilateral olfactory sensory input enhances chemotaxis behavior. *Nature Neuroscience* 11:187–99.

Porter, J., et al. (2007) Mechanisms of scent-tracing in humans. *Nature Neuroscience* 10:27–9.

Chapter 5: The ecology of smell

Brodman, J. (2009) Orchid mimics honey bee alarm pheromone in order to attract hornets for pollination. *Current Biology* 19:1368–72.

Ebrahim, S. A., et al. (2015) Drosophila avoids parasitoids by sensing their semiochemicals via a dedicated olfactory circuit. *PLoS Biology* 13:e1002318.

Goulson, D., et al. (2000) The identity and function of scent marks deposited by foraging bumblebees. *Journal of Chemical Ecology* 26:2897–911.

McBride, C. S., et al. (2014) Evolution of mosquito preference for humans linked to an odorant receptor. *Nature* 515:222–7.

Mansourian, S., et al. (2018) Wild African Drosophila melanogaster are seasonal specialists on Marula fruit. *Current Biology* 28:3960–8.

Midgley, J. J., et al. (2015) Faecal mimicry by plants found in seeds that fool dung beetles. *Nature Plants* 1:15141.

Poelman, E. H., et al. (2012) Hyperparasitoids use herbivore-induced plant volatiles to locate their parasitoid host. *PLoS Biology* 10:e1001435.

Saul-Gershenz, L. S. and Millard, J. G. (2006) Phoretic nest parasites use sexual deception to obtain transport to their host's nest. *Proceedings of the National Academy of Sciences USA* 103:14039–44.

Schiestl, F. P., et al. (1999) Orchid pollination by sexual swindle. *Nature* 399:421.

Stibor, H. and Luning, J. (1994) Predator-induced phenotypic variation in the pattern of growth and reproduction in Daphnia hyalina (Crustacea: Cladocera). *Functional Ecology* 8:97–101.

Chapter 6: Smell in culture

Balzac, H. de (1991) *Père Goriot* (Oxford: Oxford University Press).

Chiang, C. Y. (2004) Monterey-by-the-smell. *Pacific Historical Review* 73:183–214.

Classen, C., Howes, D., and Synnott, A. (1994) *Aroma: A Cultural History of Smell* (London: Routledge).

Cobb, M. (2013) *Eleven Days in August: The Liberation of Paris, 1944* (London: Simon and Schuster).

Jenner, M. S. R. (2011) Follow your nose? Smell, smelling, and their histories. *The American Historical Review* 116:335–51.
McKenzie, D. (1923) *Aromatics of the Soul: A Study of Smells* (London: Heinemann).
Majid, A. and Kruspe, N. (2018) Hunter-gatherer olfaction is special. *Current Biology* 28:409–13.
Majid, A., et al. (2018) Differential coding of perception in the world's languages. *Proceedings of the National Academy of Sciences USA* 115:11369–76.
Millu, L. (1998) *Smoke Over Birkenau* (Evanston, Ill.: Northwestern University Press).
Reinarz, R. (2014) *Past Scents: Historical Perspectives on Smell* (Urbana: University of Illinois Press).
Sutherland, J. (2016) *Orwell's Nose: A Pathological Biography* (London: Reaktion Books).

Chapter 7: The smell of the future

Bruce, T. J. A., et al. (2015) The first crop plant genetically engineered to release an insect pheromone for defence. *Scientific Reports* 5:11183.
Girling, R., et al. (2013) Diesel exhaust rapidly degrades floral odours used by honeybees. *Scientific Reports* 3:2779.
Hoover, K. C. (2018) Sensory disruption in modern living and the emergence of sensory inequities. *Yale Journal of Biology and Medicine* 91:53–62.
Jackson, D. D. (2011) Scents of place: the dysplacement of a First Nations community in Canada. *American Anthropologist* 113:606–18.
Persaud, K. (2017) Towards bionic noses. *Sensor Review* 37:165–71.
Sorokowska, A., et al. (2017) Effects of olfactory training: a meta-analysis. *Rhinology* 55:17–26.
Sunday, J. M., et al. (2014) Evolution in an acidifying ocean. *Trends in Ecology & Evolution* 29:117–25.

Further reading

Popular books about smell

Gilbert, A. (2009) *What the Nose Knows: The Science of Scent in Everyday Life* (New York: Crown).

Pelosi, P. (2016) *On the Scent: A Journey Through the Science of Smell* (Oxford: Oxford University Press).

General background for all material on mammals

Doty, R. L. (ed.) (2015) *Handbook of Olfaction and Gustation*, Third Edition (London: Wiley).

Silva Teixeira, C. S., et al. (2016) Unravelling the olfactory sense: from the gene to odor perception. *Chemical Senses* 41:105–21.

Chapter 1: How we smell

Keller, A. and Vosshall, L. V. (2016) Olfactory perception of chemically diverse molecules. *BMC Neuroscience* 17:55.

UK Science Museum video on the Chinese incense clock: <https://t.co/Wc5YzLGZry>.

Chapter 2: Smelling with genes

Keller, A., et al. (2007) Genetic variation in a human odorant receptor alters odour perception. *Nature* 449:468–72.

Chapter 3: Smell signals

Doty, R. (2010) *The Great Pheromone Myth* (Baltimore: Johns Hopkins University Press).

Wyatt, T. D. (2014) *Pheromones and Animal Behavior: Chemical Signals and Signatures* (Cambridge: Cambridge University Press).

Chapter 4: Smell, location, and memory

Wilson, D. A. and Stevenson, R. J. (2006) *Learning to Smell: Olfactory Perception from Neurobiology to Behaviour* (Baltimore: Johns Hopkins University Press).

Chapter 5: The ecology of smell

Jacobsen, R. (2019) Ghost flowers. *Scientific American* 320(2):30–9.

Chapter 6: Smell in culture

Betts, E. (ed.) (2017) *Senses of the Empire: Multisensory Approaches to Roman Culture* (Abingdon: Routledge).

Classen, C., Howes, D., and Synnott, A. (1994) *Aroma: A Cultural History of Smell* (London: Routledge).

Henshaw, V. (2013) *Urban Smellscapes: Understanding and Designing City Smell Environments* (London: Routledge).

Henshaw, V., et al. (eds) (2018) *Designing with Smell: Practices, Techniques and Challenges* (London: Routledge).

Reinarz, R. (2014) *Past Scents: Historical Perspectives on Smell* (Urbana: University of Illinois Press).

For more on smell in cinema: <http://www.cabinetmagazine.org/issues/64/turner.php>.

Chapter 7: The smell of the future

Barwich, A.-S. (2020) *Smellosophy: What the Nose tells the Mind* (Cambridge, Mass.: Harvard University Press).

Keller, A. (2016) *Philosophy of Olfactory Perception* (London: Palgrave).

Thomas, L. (1985) On smell. In: X. J. Kennedy and D. M. Kennedy (eds), *The Bedford Reader* (New York: St Martin's Press).

Index

For the benefit of digital users, indexed terms that span two pages (e.g., 52–53) may, on occasion, appear on only one of those pages.

C

D

E

F

G

H

I

Q

R

S

STATISTICS

A Very Short Introduction

David J. Hand

Modern statistics is very different from the dry and dusty discipline of the popular imagination. In its place is an exciting subject which uses deep theory and powerful software tools to shed light and enable understanding. And it sheds this light on all aspects of our lives, enabling astronomers to explore the origins of the universe, archaeologists to investigate ancient civilisations, governments to understand how to benefit and improve society, and businesses to learn how best to provide goods and services. Aimed at readers with no prior mathematical knowledge, this *Very Short Introduction* explores and explains how statistics work, and how we can decipher them.

www.oup.com/vsi

PLANETS

A Very Short Introduction

David A. Rothery

This *Very Short Introduction* looks deep into space and describes the worlds that make up our Solar System: terrestrial planets, giant planets, dwarf planets and various other objects such as satellites (moons), asteroids and Trans-Neptunian objects. It considers how our knowledge has advanced over the centuries, and how it has expanded at a growing rate in recent years. David A. Rothery gives an overview of the origin, nature, and evolution of our Solar System, including the controversial issues of what qualifies as a planet, and what conditions are required for a planetary body to be habitable by life. He looks at rocky planets and the Moon, giant planets and their satellites, and how the surfaces have been sculpted by geology, weather, and impacts.

"The writing style is exceptionally clear and pricise"

Astronomy Now

www.oup.com/vsi

PRIVACY
A Very Short Introduction
Raymond Wacks

Professor Raymond Wacks is a leading international expert on privacy. For more than three decades he has published numerous books and articles on this controversial subject. Privacy is a fundamental value that is under attack from several quarters. Electronic surveillance, biometrics, CCTV, ID cards, RFID codes, online security, the monitoring of employees, the uses and misuses of DNA, - to name but a few - all raise fundamental questions about our right to privacy. This *Very Short Introduction* also analyzes the tension between free speech and privacy generated by intrusive journalism, photography, and gratuitous disclosures by the media of the private lives of celebrities. Professor Wacks concludes this stimulating introduction by considering the future of privacy in our society.

SLEEP
A Very Short Introduction

Russell G. Foster & Steven W. Lockley

Why do we need sleep? What happens when we don't get enough? From the biology and psychology of sleep and the history of sleep in science, art, and literature; to the impact of a 24/7 society and the role of society in causing sleep disruption, this *Very Short Introduction* addresses the biological and psychological aspects of sleep, providing a basic understanding of what sleep is and how it is measured, looking at sleep through the human lifespan and the causes and consequences of major sleep disorders. Russell G. Foster and Steven W. Lockley go on to consider the impact of modern society, examining the relationship between sleep and work hours, and the impact of our modern lifestyle.

www.oup.com/vsi

SEXUALITY
A Very Short Introduction
Veronique Mottier

What shapes our sexuality? Is it a product of our genes, or of society, culture, and politics? How have concepts of sexuality and sexual norms changed over time? How have feminist theories, religion, and HIV/AIDS affected our attitudes to sex? Focusing on the social, political, and psychological aspects of sexuality, this *Very Short Introduction* examines these questions and many more, exploring what shapes our sexuality, and how our attitudes to sex have in turn shaped the wider world. Revealing how our assumptions about what is 'normal' in sexuality have, in reality, varied widely across time and place, this book tackles the major topics and controversies that still confront us when issues of sex and sexuality are discussed: from sex education, HIV/AIDS, and eugenics, to religious doctrine, gay rights, and feminism.

www.oup.com/vsi

SCIENCE AND RELIGION

A Very Short Introduction

Thomas Dixon

The debate between science and religion is never out of the news: emotions run high, fuelled by polemical bestsellers and, at the other end of the spectrum, high-profile campaigns to teach 'Intelligent Design' in schools. Yet there is much more to the debate than the clash of these extremes. As Thomas Dixon shows in this balanced and thought-provoking introduction, many have seen harmony rather than conflict between faith and science. He explores not only the key philosophical questions that underlie the debate, but also the social, political, and ethical contexts that have made 'science and religion' such a fraught and interesting topic in the modern world, offering perspectives from non-Christian religions and examples from across the physical, biological, and social sciences.

> 'A rich introductory text . . . on the study of relations of science and religion.'
>
> **R. P. Whaite, Metascience**

www.oup.com/vsi

Autism

A Very Short Introduction

Uta Frith

This *Very Short Introduction* offers a clear statement on what is currently known about autism and Asperger syndrome. Explaining the vast array of different conditions that hide behind these two labels, and looking at symptoms from the full spectrum of autistic disorders, it explores the possible causes for the apparent rise in autism and also evaluates the links with neuroscience, psychology, brain development, genetics, and environmental causes including MMR and Thimerosal. This short, authoritative, and accessible book also explores the psychology behind social impairment and savantism and sheds light on what it is like to live inside the mind of the sufferer.

www.oup.com/vsi